PRACTICAL ST

Publications from the Centre for Statistical Education

For pupils aged 10–13

Experiments in Probability	16 pages A4 size
Games Fair or Foul	9 pages A4 size
Larger or Smaller	17 pages A4 size
Shopping Baskets	25 pages A5 size

For pupils aged 13–16

Classroom Practicals	29 pages A5 size

For pupils aged 11–19

Stem and Leaf	23 pages A5 size
Box Plots	19 pages A5 size

From the Statistical Education Project 16–19

Who Goes Fast	19 pages A5 size
Learning New Skills	23 pages A5 size
Canteen Choice	29 pages A5 size
Limb Dominance	21 pages A5 size
Growing Up	30 pages A5 size
Bracken	25 pages A5 size
Looking at Data	23 pages A5 size
Mendelian Genetics	25 pages A5 size
Sales Forecasting	30 pages A5 size
Newspaper Data	42 pages A5 size
Bottles and Things	35 pages A5 size

Audio tapes for pupils and students aged over 16

The Statistician at Work	Set of boxed cassette tapes (2) with descriptive booklet

For teachers

Statistical Needs of Non-specialist Young Workers	32 pages A5 size
Practical and Project Work in A-Level Statistics	22 pages A5 size
The Best of Teaching Statistics (giving the best articles from the first five years of *Teaching Statistics*) Published by the Teaching Statistics Trust	200 pages

PRACTICAL STATISTICS

Mary Rouncefield
and
Peter Holmes

Centre for Statistical Education
Sheffield

MACMILLAN

First published 1989 by
THE MACMILLAN PRESS LTD
Houndmills, Basingstoke, Hampshire RG21 2XS
and London
Companies and representatives
throughout the world

ISBN 0-333-47344-2

A catalogue record for this book is available
from the British Library.

Reprinted 1991 (twice), 1993

Printed in Hong Kong

Contents

Preface

For many years now there has been a growing concern that statistics courses in the sixth form and at college, particularly A-level, have not been as practically oriented as they should be. Several examining boards have made attempts in different ways to improve this. The Joint Matriculation Board has an A-level Statistics syllabus which requires candidates to carry out a major project. The University of London School Examinations Board for its syllabus in A-level Pure Mathematics with Statistics requires that candidates should carry out three minor and at least two major projects relating to the topics to be examined. The University of Oxford Delegacy of Local Examinations requires its candidates for A-level Further Mathematics to have carried out different statistical practicals. Most recently the different boards' AS-level Statistics syllabuses have nearly all required candidates to carry out projects as part of their coursework and assessment. This all follows on naturally from the more coursework-oriented nature of the GCSE syllabuses and, of course, work involving statistics has been a component of the practical experimentation and fieldwork in geography, biology and psychology courses for many years.

This emphasis on practical work has posed two main problems for teachers and lecturers. The first is where to find ideas for simple practical work and the second is to find time to incorporate the work into a busy teaching programme. This book aims to meet both these needs. It gives a series of simple practical experiments that can easily be done by students, many of them in the classroom. It also links these practicals very closely with the syllabus topics so that the theory can be taught from practice. By its very nature and layout it can be used in schools, colleges of further education, and elementary courses in polytechnics. The material can easily be incorporated into many different courses including modular courses and general studies courses. Ideas and advice on the planning and execution of the practical work are provided in a separate booklet entitled *Practical Statistics: A Teachers' Guide to the Course* (ISBN 0–333–51561–7) also published by Macmillan.

The chapters have been linked to specific syllabus topics so that teachers can choose those items which are in their syllabus and make up their own practical course. This enables the book to be used by teachers of many different subjects that have a statistical content. To save time on routine drawing up of tables for recording and to make it unnecessary for students to write in this book, there are a number of photocopiable pages given at the back of the book. These pages include outline tables, populations for sampling and sheets for experiments. Copies of these pages may be made for use by students within institutions using this book, without breaching copyright. These pages are denoted by PC followed by the page number.

This book is the fruits of a 2-year project based at the Centre for Statistical Education in Sheffield. The project ran from January 1986 to December 1987 and was funded by the Department of Education and Science and we would like to express our thanks for this support. We would also like to thank all the many students, teachers, lecturers and others who helped so much with the testing and evaluation of the project's materials. The helpful and constructive comments have been taken into account in this revised published version and we are convinced that the material will be suitable and useful for many courses involving statistics.

Sheffield, 1989   M.R.
                                                                                                                            P.H.

Chapter 1 Basic Statistical Techniques

Introduction

Depending on the context in which it is used, the word 'statistics' may refer to the subject 'statistics' or to collections of sets of figures or data. A statistic is a calculation based on a set of data. As a subject, statistics is used to explain and simplify a set of figures, making them easier to understand and enabling conclusions to be drawn. So, in statistics, we draw out patterns in the data, rather than obscuring them, and we do not make data more complicated or difficult to understand.

In this chapter, you will find statistical calculations on sets of data to help to compare two (or more) sets of data and to start to draw conclusions about the two groups from which the data were collected. This chapter is an introduction to the basic techniques of data collection and simple statistics fundamental to the presentation of the results from any statistical investigation.

We present you with a choice of problems together with ideas for data collection. As you work through the techniques in this chapter, you should begin to find some useful answers and insights into your original problem. Statistics gives us useful tools to help to organise information from complex situations, to answer practical problems and ultimately to provide the basis for making decisions.

Keep carefully your data and results from this and subsequent chapters, as you will be able to make further use of them later.

Section 1.1 Practicals

Choose one problem from Group A and one from Group B. Your results and all later follow-up work for your two problems should be written up on separate sheets of paper.

Group A

Practical 1.1 Newspaper Crosswords

Problem: How do the lengths of words used in one newspaper crossword compare with those used in another newspaper crossword?

Data Collection

Find about five crosswords from each of two newspapers or magazine publications. We chose to compare crosswords in *The Guardian* and the *Evening News*. Record the lengths of the solutions straight onto a table such as Table 1.1 (using a separate table for each newspaper). Record each result using a tally mark. When you have five results together for a particular word length, the fifth tally mark should cross the previous four, making a group thus: ⦀. This will make it easier to count up your totals at the end (the total for each number of letters is entered in the column labelled Frequency).

Table 1.1   Lengths of crossword solutions

Number of letters	Tally	Frequency
1 2 3 4 5 6 7 8 9 10 11 12 13		
	Total number of words	

Practical 1.2   Delivery of Letters

Problem: How long do first-class and second-class letters take to be delivered?

Data Collection

You will require 50–100 letters in each category. Your school secretary may be willing to provide the *empty* envelopes from each day's post; alternatively you may have to collect them among your class mates. Ensure that you write the delivery date on each envelope, if your collection has to span several days. Then, by referring to the postmark, you can calculate the number of days that each letter has been in the postal system. We decided not to count Sundays in our calculations, as there are no collections or deliveries on that day.

Record your data using two tables such as Tables 1.2 and 1.3.

Table 1.2   First-class letters

Number of days	Tally	Frequency
1 2 3 4 5 6 7 8 9 10		
Total number of letters		

Table 1.3   Second-class letters

Number of days	Tally	Frequency
1 2 3 4 5 6 7 8 9 10		
Total number of letters		

Group **B**

Note that, for each of the problems in Group B, you can expect a wide range of replies or results. At this stage, we do not know what the range of replies may be, nor how to group them. Do not try to organise your results into a tally chart until later; just write down each result individually.

Practical **1.3   Earnings of Sixth-formers**

Problem: **Do male and female sixth-formers earn similar amounts from part-time employment?**

Data Collection

Ask about 100 sixth-form students how much they earn from their part-time job. If their wages vary from week to week, ask them to tell you their wage last week (which should be a week in school term time). If they do not have a job, record the answer as zero. Record males and females separately.

Practical **1.4   Playing Times of Pop Singles**

Problem: **How do the playing times compare for 'A' sides and 'B' sides of pop singles?**

Data Collection

Play at least 30 singles, recording the playing times for 'A' sides and 'B' sides separately. Do not include singles with double 'A' sides. Write down the playing time to the nearest second. This task can be shared among the members of the class and your results 'pooled' together.

Practical **1.5   Body Temperatures**

Problem: **Does everyone have the same body temperature? Do males and females have similar temperatures? Are females warmer than males? Are their temperatures more variable?**

Data Collection

Using clinical thermometers (sterilised after each person) take the temperatures of 50–100 people. Most clinical thermometers now measure in degrees Celsius and will require readings to be taken to the nearest 0.1 °C.

Section **1.2   Our Results**

Here are our results from Practical 1.2.

We collected the results shown in Table 1.4 from 98 first-class letters.

We recorded the delivery times of 104 second-class letters (Table 1.5).

Our results for Practical 1.5 appear in Table 1.6.

These data were collected for a Midlands school in July 1982 as part of the Statistical Education Project 16–19.

Looking at the Data

Your results from Group A practicals have already been organised into frequency tables, and those from Group B have not yet been organised; there are also other important differences in the data.

Table 1.4 Number of working days taken for the delivery of first-class letters

Number of working days	Frequency
1	78
2	9
3	2
4	1
5	0
6	0
7	6
8	1
9	1
10	0
Total	98

Table 1.5 Number of working days taken for the delivery of second-class letters

Number of working days	Frequency
1	8
2	55
3	31
4	8
5	0
6	0
7	0
8	1
9	0
10	1
Total	104

Table 1.6 Body temperatures

Sex	Body temperature (°C)	Sex (continued)	Body temperature (°C) (continued)	Sex (continued)	Body temperature (°C) (continued)
F	36.9	F	36.9	F	36.9
F	36.8	F	36.9	F	36.8
F	36.8	F	36.9	F	37.2
F	37.0	F	36.9	M	36.9
F	37.0	F	37.0	M	37.0
M	36.7	M	36.9	F	36.6
M	37.0	M	36.8	F	36.6
M	36.9	M	36.4	F	36.6
M	37.0	M	37.0	F	36.4
M	36.9	M	37.0	F	36.9
F	36.6	F	37.0	F	36.6
F	36.7	F	36.7	F	37.0
F	36.5	F	36.5	F	37.0
F	36.9	F	37.0	M	37.0
M	36.8	F	37.8	M	36.9
M	36.7	M	36.2	M	37.0
M	36.9	M	36.7	M	37.0
M	36.9	M	36.7	M	37.0
M	36.8	M	36.9	M	36.9
M	37.0	F	36.8	M	36.0
F	37.7	F	36.0	M	36.9
M	36.4	M	36.5	M	37.0
M	36.5	M	37.0	F	37.0
M	36.7				

F, female; M, male.

Looking at the results from Group A practicals, you will see that only integer (simple whole number) answers are possible. This would also have been true if we had been recording any of the following.

1. The number of rooms in a dwelling.
2. The number of cars passing an observation point during time intervals of 1 min.
3. The number of days taken for a letter to be delivered (measured in complete days).
4. The number of phone calls coming into a switchboard in intervals of 5 min.

Data of this type are called *discrete* data and our results can only take a finite number of possible values. (More exactly they could at most be countably infinite as in finding how many times you have to throw a die before getting a 6. There is no theoretical upper limit to the number of throws that may be needed.)

If, however, we had been finding the playing times of pop singles, our results would be more varied and might take any value within a given range. This type of data is called *continuous* data.

Occasionally, continuous data may appear to be discrete data especially if results are rounded to the nearest whole unit. If you tried Practical 1.4, the accuracy of your results will be limited by the accuracy of the timing equipment that you have available. With specialised equipment, it would be possible to find the playing time to the nearest 1/10 or 1/100 s. You have recorded your answers to the nearest second, but two discs recorded as 2 min 30 s will almost certainly not have played for exactly the same lengths of time. The recording of continuous data always requires some compromise and some decision as to what degree of accuracy is really needed. Similar problems would have been encountered if we had measured any of the following.

1. Heights of students.
2. Weights of students.
3. Distances that students have to travel to school or college.
4. Ages of students (note that ages are nearly always recorded by rounding down, and not to the nearest unit as with other continuous measures).

Since our unit of money (1 p) is so small compared with the amounts involved in recording earnings (say), we tend to treat money as though it were a continuous variable rounded to the nearest penny.

Organisation of Data

The organisation of our data for Group A practicals has already been done, during the data collection process. These results are easier to understand than those for Group B which were not recorded in an organised way. We now need to find a similar way of ordering our results from Group B (Practical 1.3, 1.4 or 1.5). A few results fall on the same value, we need to group them in some way.

Find the largest and smallest values in your data. The difference between them is called the *range*. For our temperatures (Practical 1.5) the lowest is 36.2 °C and the highest is 37.8 °C. Ideally, 6–10 groups or class intervals should be chosen across the range. We do not want so many groups that the data appear fragmented, nor too few so that the shape of the distribution is obscured.

Our data have been grouped in units of 0.2 °C giving us 10 groups with the following class intervals:

36.0 and 36.1 °C
36.2 and 36.3 °C
36.4 and 36.5 °C and so on

Table 1.7 is a frequency table drawn up for the temperatures of male and female pupils together (you might like to reanalyse them separately).

Table 1.7   A frequency table showing pupils' temperatures

Temperature (°C)	Tally	Frequency
36.0–36.1	//	2
36.2–36.3	/	1
36.4–36.5	⃫ //	7
36.6–36.7	⃫ ⃫ /	11
36.8–36.9	⃫ ⃫ ⃫ ⃫ ⃫ //	27
37.0–37.1	⃫ ⃫ ⃫ ⃫	19
37.2–37.3	/	1
37.4–37.5		0
37.6–37.7	/	1
37.8–37.9	/	1
Total number of pupils		**70**

Practical Follow-up **1.1**

1. Find the range of your results for your Group B study (Practical 1.3, 1.4 or 1.5).
2. Decide on suitable class intervals for organising your data. These class intervals should be chosen sensibly and be easy to understand.
3. Draw up separate frequency tables for your two groups (males and females, or 'A' sides and 'B' sides). Do the two groups appear similar at this stage?

Keep your follow-up work for Practicals 1.3, 1.4 or 1.5 (Group B practicals) quite separate from follow-ups for Group A practicals.

In Sections 1.3–1.8 the graphical representation of data is discussed.

Section **1.3   Line Graphs**

A line graph is used to illustrate *discrete* data. The height of each line represents the frequency. Sometimes a bar graph may be used to illustrate discrete data (or qualitative data such as the colour of eyes). In that case, the width of the bar is of no significance and again it is the height of the bar which represents the frequency. Of the two, a line graph is better, as it shows quite clearly that only integer values are possible. (There must be distinct gaps between the bars on a bar graph.)

Figures 1.1 and 1.2 are two line graphs drawn to illustrate our results for Practical 1.1. The line diagrams show very clearly the differences between the two distributions. Both are positively skewed, the distribution for first-class letters extremely so (i.e. they are not symmetrical and both have high frequencies at the left-hand end of the distribution and a long 'tail' to the right). The majority of first-class letters are delivered on the day after posting, but for this sample this was true only for 80%. Second-class letters tend to arrive on the second working day or later after posting.

Practical Follow-up **1.2**

Draw line graphs to illustrate your Group A practical results. Compare the two distributions.

Section **1.4   Histograms**

A histogram is used to illustrate continuous data (or discrete data which have been grouped). There are no gaps between the class intervals and these should cover the entire range of

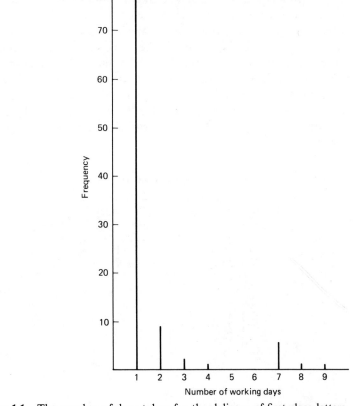

Figure 1.1   The number of days taken for the delivery of first-class letters

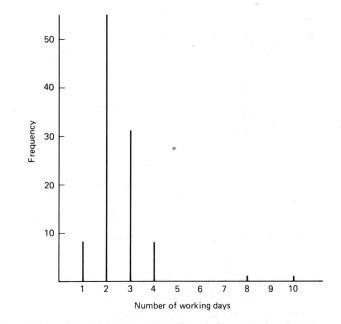

Figure 1.2   The number of days taken for the delivery of second-class letters

results. The blocks in the histogram take the width of each class interval and it is the *area* of the block which represents the frequency.

Table 1.8 again shows the temperatures of the group of 70 pupils. Note that the third column labelled *True class interval* takes account of the fact that the temperatures have been recorded to the nearest 0.1 °C. Thus a temperature measured at 36.0 °C may in fact be any value in the range from 35.95 °C to just below 36.05 °C. Whenever continuous data are being used, the problem of accuracy of measurement must be taken into account. The class intervals are all of equal width (0.2 °C) and so the blocks in the histogram will be of equal width also (Figure 1.3). (In this situation the frequency will in fact be proportional to the height.)

Table 1.8   Body temperatures of a group of 70 pupils

Temperature (°C)	Frequency	True class interval
36.0–36.1	2	$35.95 \leqslant x < 36.15$
36.2–36.3	1	$36.15 \leqslant x < 36.35$
36.4–36.5	7	$36.35 \leqslant x < 36.55$
36.6–36.7	11	$36.55 \leqslant x < 36.75$
36.8–36.9	27	$36.75 \leqslant x < 36.95$
37.0–37.1	19	$36.95 \leqslant x < 37.15$
37.2–37.3	1	$37.15 \leqslant x < 37.35$
37.4–37.5	0	$37.35 \leqslant x < 37.55$
37.6–37.7	1	$37.55 \leqslant x < 37.75$
37.8–37.9	1	$37.75 \leqslant x < 37.95$

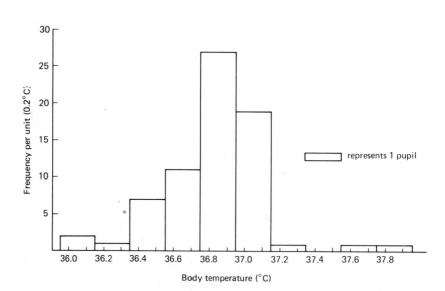

Figure 1.3   Histogram of the body temperatures of 70 pupils corresponding to Table 1.8

As there are very few pupils with temperatures below 36.4 °C or above 37.4 °C, the information in Table 1.8 could be reorganised as in Table 1.9. Here the first interval covers temperatures in the range from 35.95 °C up to 36.35 °C (but not including 36.35 °C) and has a width of 0.4 °C. Similarly the last interval includes temperatures from 37.35 °C to

Table 1.9 Body temperatures of a group of 70 pupils

Temperature (°C)	Frequency	True class interval
36.0–36.3	3	$35.95 \leqslant x < 36.35$
36.4–36.5	7	$36.35 \leqslant x < 36.55$
36.6–36.7	11	$36.55 \leqslant x < 36.75$
36.8–36.9	27	$36.75 \leqslant x < 36.95$
37.0–37.1	19	$36.95 \leqslant x < 37.15$
37.2–37.3	1	$37.15 \leqslant x < 37.35$
37.4 and above	2	$37.35 \leqslant x < 37.95$

just below 37.95 °C and has a width of 0.6 °C. If a particular class is 2 or 3 times the standard width, then the height of the block must be adjusted accordingly to 1/2 or 1/3 of the frequency, so that the *area* is in the correct proportion (Figure 1.4). Note that the vertical axis is labelled Frequency per unit (0.2 °C) as it is the area (and not the height) of each block which shows the frequency. The boundary lines for each block are at the limits of the true class intervals.

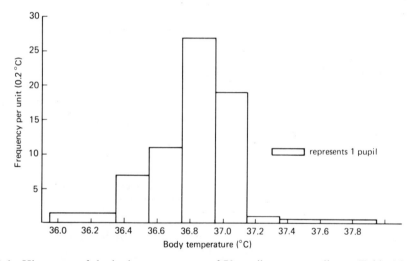

Figure 1.4 Histogram of the body temperatures of 70 pupils corresponding to Table 1.9

Practical Follow-up **1.3**

Draw histograms with, firstly, equal class intervals and, secondly, unequal class intervals to illustrate your results to Practical 1.3, 1.4 or 1.5.

Section **1.5** A Frequency Polygon

A frequency polygon is used to illustrate *continuous* data. It is obtained by plotting the frequencies at the central point in each class interval and joining the points with straight lines. If any of the class intervals are wider than the others, the height of the point should be adjusted in the same way as for a histogram. It may be easier to imagine that you are joining the midpoints of the tops of the blocks in the underlying histogram. Figure 1.5 is a frequency polygon based on the histogram shown in Figure 1.3.

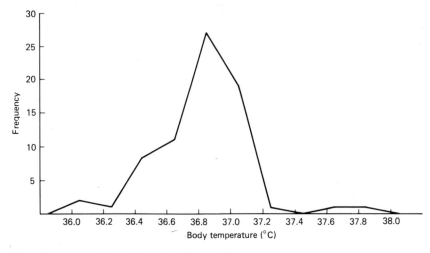

Figure 1.5 Frequency polygon based on Figure 1.3

***Practical Follow-up* 1.4**

Draw frequency polygons based on your histograms drawn for Practical follow-up 1.3.

Section **1.6** **Stem-and-leaf Display**

The results used in this section were collected in a local sixth-form college where students were asked what their part-time earnings were in a typical week (Table 1.10).

Table 1.10 Part-time earnings for sixth-form students

Part-time earnings for males (£)							
0	0	4.50	15.00	0	0	12.50	12.50
0	17.50	25.00	0	19.30	10.00	18.75	5.00
15.00	8.50	18.20	16.90	16.50	17.00	7.00	14.25
35.00	0	0	20.50	0	19.50	18.05	7.50
0	0	9.50	0	3.50	0	13.50	24.50
0	0	4.25	22.50	27.00			
Part-time earnings for females (£)							
0	0	8.00	12.50	14.50	0	0	0
17.50	16.05	0	20.00	0	26.50	4.00	11.50
5.00	10.00	16.00	0	25.00	0	14.50	0
8.50	7.50	22.75	3.75	15.00	21.75	17.50	9.00
18.00	21.40	5.25	14.75	0	6.00	0	10.68
18.20	14.20	13.70	0	20.00	0	9.75	21.50
7.50	19.00	0	0	8.25	11.75	8.50	16.00
12.80	0	22.75	14.25	13.50			

A stem-and-leaf display has the advantages of a bar graph or block diagram, while also retaining much of the detail of the original data.

We have first constructed a stem-and-leaf display directly from the data for male students' part-time earnings. Usually only two significant figures can be shown: so we have recorded earnings to the nearest £. Each row in the display is for a different 'tens' digit, and the units digit for any figure is written in the appropriate row.

Our first four results were 0, 0, £4.50 (rounded up to £5) and £15. This is how these results are recorded:

	0,	0,	£5,	£15

(tens)	(units)		
0	0	0	5
1	5		
2			
3			

1|5 means £15

The display in Table 1.11 shows the data in each category recorded in the order in which the results were collected. If you prefer, a new display may be drawn up, showing the results in numerical order (Table 1.12). (This can be helpful later for identifying medians, etc.)

Table 1.11   Male students' part-time earnings in the order that the data was collected·

(tens)	(units)
0	0 0 5 0 0 0 0 5 9 7 0 0 0 8 0 0 0 4 0 0 0 4
1	5 3 3 8 9 0 9 5 8 7 7 7 4 8 0 4
2	5 1 0 5 3 7
3	5

3|5 means £35

Table 1.12   Male students' part-time earnings in numerical order

0	0 0 0 0 0 0 0 0 0 0 0 0 0 0 0 0 4 4 5 5 7 8 9
1	0 0 3 3 4 4 5 5 7 7 7 8 8 8 9 9
2	0 1 3 5 5 7
3	5

3|5 means £35

If there are large frequencies in each class, more detail may be given by splitting each 'ten' into two rows (Table 1.13). The top row is used for units 0–4 and the second for units 5–9.

Table 1.13   Stem-and-leaf display of male students' part-time earnings

0	0 0 0 0 0 0 0 0 0 0 0 0 0 0 0 4 4
0	5 5 7 8 9
1	0 0 3 3 4 4
1	5 5 7 7 7 8 8 8 9 9
2	0 1 3
2	5 5 7
3	
·3	5

3|5 means £35

Table 1.14 Sixth-formers weekly part-time earnings: a stem-and-leaf display

Females	Males

```
4 4 0 0 0 0 0 0 0 0 0 0 0 0 0 0 0 0 0 0 | 0 | 0 0 0 0 0 0 0 0 0 0 0 0 0 0 0 0 4 4
              9 9 9 8 8 8 6 6 5 5 | 0 | 5 5 7 8 9
              4 4 4 4 3 3 2 2 1 0 0 | 1 | 0 0 3 3 4 4
        9 8 8 8 8 6 6 6 5 5 5 5 | 1 | 5 5 7 7 7 8 8 8 9 9
                    3 3 2 2 1 0 0 | 2 | 0 1 3
                            8 7 5 | 2 | 5 5 7
                                  | 3 |
                                  | 3 | 5
```

3|5 means £35

Stem-and-leaf displays are particularly useful for the direct comparison of two distributions, by placing them 'back to back'. Table 1.14 shows a comparison of the part-time earnings of male and female sixth-formers. This display shows very clearly the similarity of the two sets of figures.

Practical Follow-up **1.5**

Draw stem-and-leaf displays to illustrate your data from your Group B practical (Practical 1.3, 1.4 or 1.5).
    What do they show you?

Section **1.7** **Pie Charts**

A pie chart may be drawn to illustrate discrete or continuous data, and is especially suitable for qualitative data—colour of eyes, voting behaviour, method of travelling to work, etc. It is a circular diagram divided proportionally to show the various subgroups within a total population. It is ideal, therefore, for showing percentages and proportions but may be used for frequency tables. The categories must be mutually exclusive and their sum must be a meaningful 100%.
    Table 1.15 is again our table of results for the part-time earnings of male sixth-formers. You will notice that the first class interval has an especially high frequency. This is because it contains all the people who do not work at all and have zero earnings. It would be much better to take these 15 students out to form a separate group. If we do that, however, the

Table 1.15 Part-time earnings of male sixth-formers

Part-time earnings (£)	Frequency
0– 4.99	18
5.00– 9.99	5
10.00–14.99	5
15.00–19.99	11
20.00–24.99	3
25.00–29.99	2
30.00–34.99	0
35.00–39.99	1
Total	45

Table 1.16 Rearrangement of Table 1.15

Part-time earnings (£)	Frequency
0	15
0.01– 5.00	4
5.01–10.00	5
10.01–15.00	6
15.01–20.00	9
20.01–25.00	4
25.01–30.00	1
30.01–35.00	1
Total	45

class intervals will have to be rearranged into slightly different classes (to keep them of equal widths) and the raw data reclassified. Table 1.16 is the new table for male sixth-formers.

The group who do not work form 1/3 of the total group, and we shall get a clearer picture if they can be shown separately (rather than mixed in with the low-earning workers who earn less than £5.00) (Table 1.17). Unfortunately, it is then impossible to show a group at zero on a histogram. However, a pie chart does not have the problem as we have no horizontal scale, and we shall be able to show clearly that 1/3 of our group of male sixth-formers do not have paid work. Some of the other classes have been amalgamated to make the diagram simpler (but this is not strictly necessary).

Table 1.17 Part-time earnings of male sixth-formers

Earnings (£)	Frequency	Relative frequency (proportion of total)	Angle in pie chart (deg)
Zero	15	$\frac{15}{45} = \frac{1}{3} = 0.3\dot{3}$	$0.3\dot{3} \times 360 \approx 120$
0.01–10.00	9	$\frac{9}{45} = \frac{1}{5} = 0.20$	$0.20 \times 360 = 72$
10.01–20.00	15	$\frac{15}{45} = \frac{1}{3} = 0.3\dot{3}$	$0.3\dot{3} \times 360 \approx 120$
Over 20.00	6	$\frac{6}{45} = \frac{2}{15} = 0.1\dot{3}$	$0.1\dot{3} \times 360 \approx 48$
Total	45	1.00	360

For each subgroup the relative frequency, or proportion of the total, is calculated. Then the number of degrees allocated in the pie chart is found by multiplying the relative frequency by 360° (Figure 1.6). The area in each section represents the proportion of the total who appear in that class. (If you have a circular protractor graded from 0 to 100, then the numbers in the third column, when multiplied by 100, give the angles to measure.)

If we also wish to illustrate the part-time earnings of female sixth-formers on a pie chart, we shall need to make the total area of their pie chart larger, in order to show that there are 61 female students, as opposed to only 45 male students.

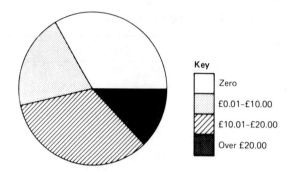

Figure 1.6   Part-time earnings of male sixth-formers

The ratio of the two areas must be 45 to 61. Therefore the radii will be in the ratio $\sqrt{45}$ to $\sqrt{61}$. Our pie chart for males had a radius of 2 cm and so the radius for females will be

$$2 \text{ cm} \times \sqrt{\frac{61}{45}} = 2.33 \text{ cm}$$

The part-time earnings for female students are shown in Table 1.18.

Table 1.18   Part-time earnings of female sixth-formers

Earnings (£)	Frequency	Relative frequency (proportion of total)	Angle in pie chart (deg)
Zero	17	$\frac{17}{61} \approx 0.279$	$0.279 \times 360 \approx 100.3$
0.01–10.00	14	$\frac{14}{61} \approx 0.230$	$0.230 \times 360 \approx 82.6$
10.01–20.00	23	$\frac{23}{61} \approx 0.377$	$0.377 \times 360 \approx 135.7$
Over 20.00	7	$\frac{7}{61} \approx 0.115$	$0.115 \times 360 \approx 41.3$
Total	61	≈ 1.0	≈ 360

Figure 1.7 shows the two pie charts drawn together. They show very clearly the similarity between the two sets of earnings with virtually the same proportions appearing in each earnings group.

Practical Follow-up **1.6**

1. Draw pie charts to compare your two (or four) sets of data for your Group A practical (Practical 1.1 or 1.2).
2. Draw pie charts to compare your two sets of data for your Group B practical (Practical 1.3, 1.4 or 1.5).

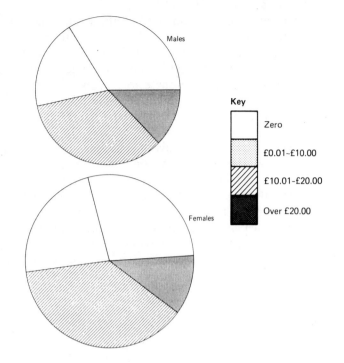

Figure 1.7   Part-time earnings of sixth-formers

Section **1.8**   **Cumulative Frequency Diagrams**

Cumulative frequency diagrams are most often used for *continuous* data (or discrete data which have been grouped).

A cumulative frequency table may be constructed by taking a running total down the frequency column (adding on each frequency to the total as you go down).

From Table 1.19, we find that 2 pupils had temperatures of less than $36.15\,^{\circ}\text{C}$, 3 had

Table 1.19   Cumulative frequency table for pupils' body temperatures

Temperature (°C)	Frequency	Upper limit for temperature (°C)	Cumulative frequency
36.0–36.1	2	$x < 36.15$	2
36.2–36.3	1	$x < 36.35$	3
36.4–36.5	7	$x < 36.55$	10
36.6–36.7	11	$x < 36.75$	21
36.8–36.9	27	$x < 36.95$	48
37.0–37.1	19	$x < 37.15$	67
37.2–37.3	1	$x < 37.35$	68
37.4–37.5	0	$x < 37.55$	68
37.6–37.7	1	$x < 37.75$	69
37.8–37.9	1	$x < 37.95$	70

temperatures of less than 36.35 °C and so on. Note that we take into account the true class intervals here, as we did when we drew histograms in Section 1.4. The first class interval has its true limits from 35.95 °C to just below 36.15 °C, and the second from 36.15 °C to just below 36.35 °C and so on. This type of consideration must always be made for continuous data, so that there are no gaps in the range of data. We must take into account the way in which measurements have been rounded (to the nearest 0.1 °C here). The cumulative frequency for a particular temperature is the frequency of observations below that value.

By plotting the cumulative frequency against the correct upper limit for each class interval, a cumulative frequency diagram is obtained (Figure 1.8). A cumulative frequency polygon is drawn by joining the points by straight lines. Alternatively the points may be joined by a smooth curve. There has been considerable discussion about the relative merits of these two methods. The cumulative frequency polygon assumes an even distribution of values within each class interval and so is consistent with the horizontal lines at the top of the blocks on the histogram. The cumulative frequency curve attempts to represent the possible non-uniform (uneven) distribution of values within each class.

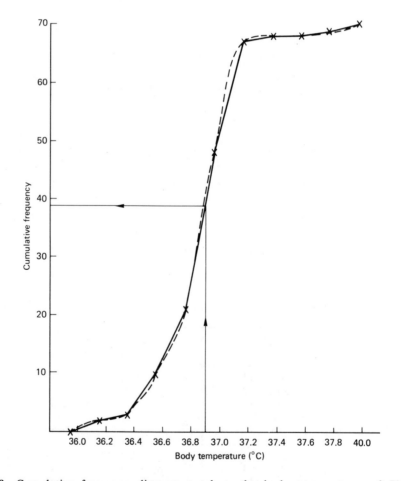

Figure 1.8 Cumulative frequency diagram to show the body temperatures of 70 pupils: ---, cumulative frequency curve; ———, cumulative frequency polygon

From a cumulative frequency diagram it is possible to answer questions of the type 'How many pupils have a temperature of less than 36.9 °C?' The answer is found by drawing a vertical line up to the graph from 36.9 °C. By reading off the cumulative frequency for that amount the answer is found to be 39.

Practical Follow-up 1.7

Draw cumulative frequency diagrams to illustrate both sets of your results for Practical 1.3, 1.4 or 1.5 (Group B practicals).

Sections 1.9–1.11 contain discussions of the measures showing the centre of a set of observations (averages).

Section 1.9   The Mode

The mode is the single value or subgroup which occurs with the highest frequency. If the data are grouped, then the modal class is that subgroup with the highest frequency.

For *discrete data* the mode may be easily extracted from the frequency table or frequency diagram. Tables 1.20 and 1.21 are our results for Practical 1.1.

Table 1.20 Frequency table for the lengths of words used in crossword puzzles in *The Guardian*

Number of letters	Frequency
3	10
4	46
5	86
6	94
7	78
8	50
9	20
10	12
11	10
12	12
13	10
Total	428

Table 1.21 Frequency table for the lengths of words used in crossword puzzles in the *Evening News*

Number of letters	Frequency
3	166
4	226
5	225
6	178
7	75
8	21
9	20
10	2
11	1
12	0
13	3
Total	917

The modal word length is 6 letters for *The Guardian* crosswords, and 4 letters for *Evening News* crosswords. If you draw line graphs to illustrate these data, you will notice that the mode is the peak of the frequency distribution.

Continuous Data

For continuous data condensed into a frequency table (Table 1.22), it will not be possible to give a single value as the mode. Instead, we shall have to give the class interval with the highest frequency as the modal group (or class). For male students (who work) the modal class for their part-time earnings is £15.01–£20.00. Also it is really sensible to talk about a modal class only when all the class intervals are equal.

Table 1.22   Part-time earnings for male students with jobs

Amount (£)	Frequency
0.01– 5.00	4
5.01–10.00	5
10.01–15.00	6
15.01–20.00	9
20.01–25.00	4
25.01–30.00	1
30.01–35.00	1

Section 1.10   The Median

Discrete Data

The median is the central value in a set of data which have been arranged in order of size.

To find the median length of words used in *The Guardian* crosswords, we first draw up a cumulative frequency table (Table 1.23). (The frequency table has already organised the data for us in order of size.)

Table 1.23   Cumulative frequency table for the lengths of words in *The Guardian* crosswords

Number of letters x	Frequency	Upper limit for x	Cumulative frequency
3	10	$x \leqslant 3$	10
4	46	$x \leqslant 4$	56
5	86	$x \leqslant 5$	142
6	94	$x \leqslant 6$	236
7	78	$x \leqslant 7$	314
8	50	$x \leqslant 8$	364
9	20	$x \leqslant 9$	384
10	12	$x \leqslant 10$	396
11	10	$x \leqslant 11$	406
12	12	$x \leqslant 12$	418
13	10	$x \leqslant 13$	428
	428		

We need now to locate the middle word and to find its length.

How to Locate the Middle Item of n *Observations*

1. Where n is odd, the middle value is at $(n + 1)/2$ (e.g. if we have 65 observations the middle one is the 33rd).
2. Where n is even, there are two middle values located at $n/2$ and $n/2 + 1$. The median is taken as the average of these values.

We have 428 words in our set; so we need to find the number of letters for the two middle words. These are at the $n/2 = 214$th word and the $n/2 + 1 = 215$th word.

To locate these words, look down the cumulative frequency column for a total just above these. The cumulative frequency that we shall take is 236; this occurs for 6 letters. Both the 214th and the 215th words have 6 letters (so their average is also 6 letters). Therefore the median word length is 6 letters for words in *The Guardian* crossword puzzles.

Practical Follow-up 1.8

1. Find the median number of letters for words used in *Evening News* crosswords (from our data in Section 1.9, page 17).
2. Find the median for each of your sets of data obtained from Group A practicals (Practicals 1.1 and 1.2).

Section 1.11   Finding the Median of Continuous Data by Linear Interpolation

There are two methods of finding the median for grouped data. The first method uses linear interpolation from a cumulative frequency table. Table 1.24 shows the weekly part-time earnings for female students, with the cumulative frequencies calculated.

Table 1.24   Weekly part-time earnings for female students

Earnings (£)	Frequency	Upper limit for earnings (£)	Cumulative frequency
0.01– 5.00	3	$x < 5.005$	3
5.01–10.00	11	$x < 10.005$	14
10.01–15.00	13	$x < 15.005$	27
15.01–20.00	10	$x < 20.005$	37
20.01–25.00	6	$x < 25.005$	43
25.01–30.00	1	$x < 30.005$	44
Total	44	Total	168

There are 44 female students who have part-time jobs; so the median value of their earnings will be midway between the 22nd and 23rd values. If we look down the cumulative frequency column, the total next above 22 and 23 is 27 which occurs for earnings up to £15.005. Therefore, we know that the median must lie between £10.005 and £15.005. We also know that there are 13 students who earn amounts in that interval. Assuming that we do not have access to the original raw data, we split the £5.00 in that class interval into 13 equal spaces (Figure 1.9). The first student in this £5.00 interval is the 15th and the last is the 27th (from the complete group of 44).

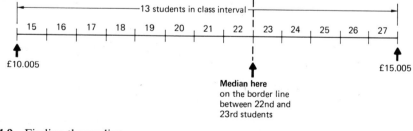

Figure 1.9   Finding the median

How to Find the Value of the Median

1. Find the width of the class interval containing the median (£5.00 in our data).
2. Find the number of observations spread across that group (i.e. the frequency for that group, which is 13 for our data).
3. Divide the interval into the required number of spaces. For our data, each space = £5.00 ÷ 13 = £0.385.
4. Draw a diagram to decide on the position of the median. For our data, it is at the end of the 22nd student's space.
5. How many spaces along the class interval, containing the median, do you need to go? Counting along, we find that we need to go 8 spaces along the interval between £10.005 and £15.005. Alternatively the median can always be found at the position where $n/2$ would be found proportionally through the appropriate class interval. Note that the 22–33 borderline is 8/13 through the interval and *not* $8\frac{1}{2}/13$ which would be equivalent to $(n+1)/2$.
6. Calculate the median thus:

median = lower bound of class interval + correct distance along
$$= £10.005 + (8 \times £0.385)$$
$$= £10.005 + £3.08$$
$$= £13.085$$
$$= £13.09 \text{ to the nearest penny}$$

The median earnings for female students (found by this method of linear interpolation) is £13.09.

Interestingly, while the distributions for male and female earnings appeared to be very similar, the median earnings for females is over £2.00 lower than the median for males.

Section 1.12 Finding the Median of Continuous Data Using a Cumulative Frequency Diagram

The other method of finding the median uses a cumulative frequency diagram (as shown in Section 1.7). Either the curve or the frequency polygon may be used.

To locate the median for n items, a horizontal line should be drawn across to the graph at $n/2$ on the cumulative frequency. This method always locates the middle of the distribution whether the total cumulative frequency n is odd or even and is consistent whether we work from the top down or from the bottom up.

We use this method below to find the median temperature for our group of 70 pupils considered in Table 1.19 (Figure 1.10).

We have 70 pupils and so the median can be found at $n/2 = 35$ on the cumulative frequency axis. By drawing a line across at this point the median temperature can be found. From the graph in Figure 1.10, we find that the median temperature is 36.85 °C.

Practical Follow-up **1.9**

Find the median for each of your sets of data for Practical 1.3 or 1.4 (Group B practicals).

Section 1.13 The Mean

The mean is the simple average of a set of observations. It is calculated by dividing the sum of the observations by the number of observations in the group.

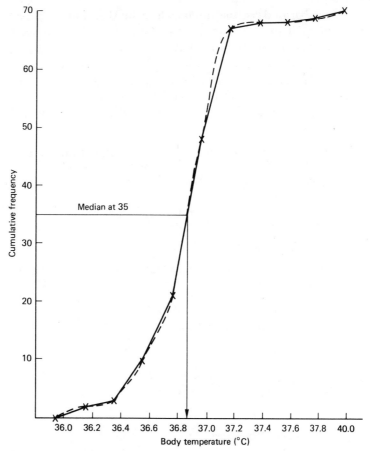

Figure 1.10 Finding the median temperature: – – –, cumulative frequency curve; ———, cumulative frequency polygon

Refer back to Table 1.5 in which we give the frequency table of the number of working days taken for second-class letters to be delivered.

To calculate the mean number of working days taken, we need to find the total number of days for all the 104 letters. Table 1.25 summarises the data for these 104 letters; for each

Table 1.25   The mean number of working days taken for second-class letters to be delivered

Number x_i of working days	Frequency f_i	$f_i x_i$
1	8	8
2	55	110
3	31	93
4	8	32
5	0	0
6	0	0
7	0	0
8	1	8
9	0	0
10	1	10
Total	104	261

number of days (the variate which we shall call x_i) a frequency is given (usually denoted by f_i). Thus, we have 8 letters which took 1 day to be delivered, 55 letters which took 2 days and so on. The total number of days taken by the letters delivered in 1 day is 8 (i.e. 8×1) and, for those taking 2 days, there is a total of 110 days (i.e. 55×2).

To find the total number of days altogether, we must multiply each number of days (or each value of x_i) by its frequency. So we need another column in the table for f_i multiplied by x_i (to be labelled $f_i x_i$). The mean will be the total of this column divided by 104.

Thus,

total number of working days = 261 (total of the $f_i x_i$ column)
number of letters   = 104
mean number of working days = 2.51 (to two decimal places)

Using this method, we find that our sample of first-class letters took, on average, 1.68 working days to be delivered.

A General Formula for the Mean

The mean of a variate x is usually denoted by the symbol $\bar{x}$.

In our example,

$$\bar{x} = \frac{\text{total number of working days}}{\text{number of letters}}$$

Generally,

$$\bar{x} = \frac{\text{sum of all the observation values}}{\text{number of observations}}$$

$$\bar{x} = \frac{\text{sum of } f_i x_i}{\text{sum of frequencies}}$$

Another symbol $\sum$ (the greek capital letter sigma) is used to denote the *sum*. Thus

$$\bar{x} = \frac{\sum_i f_i x_i}{\sum_i f_i}$$

The suffix i denotes that we have to sum over all values of i; here i represents the number of different classes (there are 11 classes in this example).

Practical Follow-up **1.10**

Calculate the mean for each set of data collected for your Group A practical (Practical 1.1 or 1.2).

Section **1.14**   **Calculating the Mean of a Grouped Frequency Table**

Refer back to Table 1.22 and look at the frequency table for the part-time earnings of our male sixth-formers, and see that each frequency is given for a range of values, e.g. £0.01–£5.00, rather than for a single value, as in the previous section.

Without the original data, it is impossible to find the precise earnings of any one individual student from this table. If this type of frequency table is the only information available, then only an estimate of the mean can be calculated.

We know that the mean is calculated as

$$\bar{x} = \frac{\sum_i f_i x_i}{\sum_i f_i}$$

and that each value of x_i must be multiplied by its frequency. As precise values of x_i are not given here, the centre of each class interval is taken as x_i (Table 1.26). (To find the central value of a class interval, add the lower and upper limits together and divide by 2.)

Table 1.26 Earnings of male sixth-formers

Earnings (£)	Middle x_i of class interval	f_i	$f_i x_i$
0.01– 5.00	£2.505	4	10.02
5.01–10.00	£7.505	5	37.525
10.01–15.00	£12.505	6	75.03
15.01–20.00	£17.505	9	157.545
20.01–25.00	£22.505	4	90.02
25.01–30.00	£27.505	1	27.505
30.01–35.00	£32.505	1	32.505
	Total	30	430.15

$$\sum_i f_i x_i = £430.15$$

$$\sum_i f_i = 30$$

$$\bar{x} = \frac{\sum_i f_i x_i}{\sum_i f_i}$$

$$\bar{x} = \frac{£430.15}{30} = £14.34$$

Thus the mean earnings for the 30 male students who have part-time jobs is £14.34.

Practical Follow-up 1.11

1. Calculate the mean earnings for female students based on our data (Table 1.27).
2. Calculate the mean for each set of data that you collected for your Group B practical (Practical 1.3, 1.4 or 1.5).

Table 1.27   Earnings for female students

Earnings (£)	Frequency
0.01– 5.00	3
5.01–10.00	11
10.01–15.00	13
15.01–20.00	10
20.01–25.00	6
25.01–30.00	1

Section **1.15   Change of Origin and Units for Grouped Data**

Table 1.28 is the frequency table showing the temperature of our group of 70 pupils.

Table 1.28   Frequency table of body temperatures of 70 pupils

Temperature (°C)	Middle x_i of class interval	Frequency f_i
36.0–36.1	36.1	2
36.2–36.3	36.3	1
36.4–36.5	36.5	7
36.6–36.7	36.7	11
36.8–36.9	36.9	27
37.0–37.1	37.1	19
37.2–37.3	37.3	1
37.4–37.5	37.5	0
37.6–37.7	37.7	1
37.8–37.9	37.9	1

To find the mean, we shall require

$$\frac{\sum_i f_i x_i}{\sum_i f_i}$$

First, each value of x_i (taking the middle of each class interval) must be multiplied by its frequency to find $f_i x_i$. Note, however, that all our x_i values are between 36 and 38 °C and that we would simplify our calculations by taking a working origin at 36 or 37 °C. If we use 37 as our working origin, we then recalculate our x_i values as d_i taken as the difference of $x_i - 37$ (Table 1.29):

$d_i = x_i - 37$

Therefore,

$x_i = d_i + 37$

Now $\bar{x}$ can be calculated as

$\bar{x} = \bar{d} + 37$

Table 1.29 Recalculation using 37 °C as the origin

Temperature (°C)	Middle x_i of class interval	$x_i - 37 = d_i$	f_i	$f_i d_i$
36.0–36.1	36.1	−0.9	2	−1.8
36.2–36.3	36.3	−0.7	1	−0.7
36.4–36.5	36.5	−0.5	7	−3.5
36.6–36.7	36.7	−0.3	11	−3.3
36.8–36.9	36.9	−0.1	27	−2.7
37.0–37.1	37.1	0.1	19	1.9
37.2–37.3	37.3	0.3	1	0.3
37.4–37.5	37.5	0.5	1	0.0
37.6–37.7	37.7	0.7	1	0.7
37.8–37.9	37.9	0.9	1	0.9
			Total	−12 +3.8

$$\bar{x} = \frac{\sum_i f_i d_i}{\sum_i f_i} + 37$$

$$\sum_i f_i d_i = -8.2$$

$$\sum_i f_i = 30$$

$$\bar{d} = \frac{-8.2}{30}$$

$$= 0.273$$

So

$$\bar{x} = 37 - 0.273$$

$$= 36.727$$

$$= 36.73\ °C$$

This method does make calculations easier, and often a change of units may be used also. In this example, we could have used units of 0.1 °C to eliminate the need for any decimals in the working. This method is not as important as it used to be now that calculators and microcomputers are widely used. However, to find the mean of 3.001, 3.002 and 3.004 (say), it is certainly much easier to find the mean of 1, 2 and 4 and then to adjust the answer.

In Sections 1.16–1.20, measures of variability within a group are discussed.

Section 1.16 The Range

One of the simplest measures of variability within a group is the range. This is calculated by subtracting the lowest observation from the highest.

We can calculate the range for the lengths of words used in *The Guardian* and *Evening News* crosswords in order to compare the two distributions.

The Guardian Crosswords

We have

lowest number of letters in a word (bottom of range) = 3
highest number of letters in a word (top of range)   = 13

Therefore,

$$\text{range} = 10 \text{ letters}$$

Evening News Crosswords

We have

lowest number of letters in a word (bottom of range) = 3
highest number of letters in a word (top of range)   = 13

Therefore,

$$\text{range} = 10 \text{ letters}$$

The two groups of crosswords have the same range for the number of letters per word, and we shall need a more sophisticated measure of variability if we are to highlight the very real differences between the two distributions.

Section 1.17   The Interquartile Range

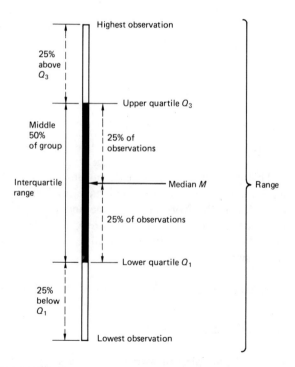

Figure 1.11   The interquartile range

The interquartile range gives the range of the central 50% of any group.

The median is the very central observation in the group, and the interquartile range extends from the value 25% through the group to that 75% through the group (Figure 1.11).

To calculate the interquartile range, we first locate the observation which is 25% of the way through the group, moving upwards from the lowest value. This observation is called the lower quartile (or first quartile Q_1). Then we locate the observation which is 75% of the way through the group, which is called the upper quartile (or third quartile Q_3). The interquartile range is found by subtracting, i.e. $Q_3 - Q_1$.

The Interquartile Range for Discrete Data

The quartiles may be found directly from a cumulative frequency table. Table 1.30 shows the data for *The Guardian* crosswords.

Table 1.30 Cumulative frequency table for *The Guardian* crosswords

Number of letters	Frequency	Cumulative frequency
3	10	10
4	46	56
5	86	⟦142⟧ ←Q_1 107th and 108th in here
6	94	236
7	78	314
8	50	⟦364⟧ ←Q_3 320th and 321st in here
9	20	384
10	12	396
11	10	406
12	12	428
13	10	
Total	428	

If we know all the individual values for our data, then the lower quartile is the reading that has the ratio of points below to points above as $1:3$. There are several different ways that are used to do this. The most consistent rule appears to be as follows.

1. If n is a multiple of 4, take the average of the $(n/4)$th and $(n/4 + 1)$th values.
2. If n is not a multiple of 4, then find the nearest integer to $(n + 2)/4$ and use this reading.

Here $n = 428$ and is a multiple of 4; so look for the mean of the 107th and 108th readings in the table. Both are found in the group of words with 5 letters.

So $Q_1 = 5$ letters.

Q_3 is located at the same distance from the top of the distribution as Q_1 is from the bottom of the distribution (i.e. the average of the 320th and 321st readings). Again, these two readings are in the same group (with 8 letters).

So

upper quartile $= 8$ letters

interquartile range $= Q_3 - Q_1$

$$= 8 - 5$$

$$= 3 \text{ letters}$$

Thus, for *The Guardian* crosswords, the middle 50% of words lie between 5 and 8 letters in length.

***Practical Follow-up* 1.12**

1. Find the median, lower quartile and upper quartile for the lengths of words used in *Evening News* crosswords. (Our data are reproduced in Table 1.21.)
2. Find the lower and upper quartiles for your sets of data from Practical 1.1 or 1.2.

Section **1.18**   **'Box-and-Whisker' Diagrams**

Comparison of *The Guardian* and *Evening News* Crosswords

'Box-and-whisker' diagrams are very useful for making a comparison of two groups. For their construction, the following values need to be found first.

1. Top of range.
2. Bottom of range.
3. Median.
4. Lower quartile.
5. Upper quartile.

The simple box-and-whisker diagram is set out as in Figure 1.12.

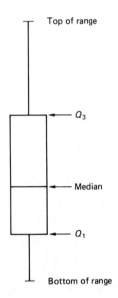

Figure 1.12   Simple box-and-whisker diagram

The interquartile range forms the 'box' while the range forms the 'whiskers'.

Figure 1.13 shows box-and-whisker diagrams for *The Guardian* and *Evening News* crosswords (drawn on the same scale). This shows quite clearly the differences between the two distributions. The bulk of the distribution for the *Evening News* is only at the same value as the median for *The Guardian*. There are many more longer words in *The Guardian* crossword puzzles, whereas over 25% of words in the *Evening News* crosswords have only 3 or 4 letters.

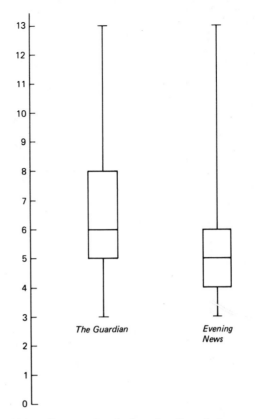

Figure 1.13   Box-and-whisker diagrams for the lengths of words in crossword puzzles

Practical Follow-up **1.13**

1. For the data from your Group A practical study (Practical 1.1 or 1.2) find the following values.
   a.   Top of range.
   b.   Bottom of range.
   c.   The median.
   d.   Upper and lower quartiles.
   You should have the answers to c and d available from previous follow-ups.
2. Draw a box-and-whisker diagram for each set of data that you have.

Section **1.19**   **Interquartile Range for Continuous or Grouped Data**

For continuous (or grouped) data the interquartile range is most easily found from a cumulative frequency diagram.

Figure 1.14 is the cumulative frequency diagram for the temperatures of the group of 70 pupils. Refer back to Table 1.19 for the cumulative frequency table.

The position of the lower quartile Q_1 is at $n/4$ for n observations. Thus, with $n = 70$, we require the value of the observation at position $70/4 = 17.5$. The value of Q_1 is found by drawing a horizontal line at 17.5 over to the graph and reading off the value on the

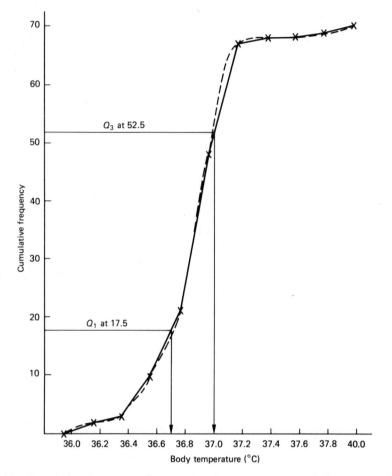

Figure 1.14 Cumulative frequency diagram for the temperatures of the group of 70 pupils: ---, cumulative frequency curve; ———, cumulative frequency polygon

horizontal axis:

lower quartile $Q_1 = 36.68\,^\circ\text{C}$

Thus, 25% of the pupils had temperatures lower than $36.68\,^\circ\text{C}$.

The upper quartile Q_3 is located at $3n/4 = 210/4 = 52.5$. This is the position of Q_3, and the value of Q_3 must be read off from the graph:

upper quartile $Q_3 = 37.0\,^\circ\text{C}$

Therefore,

interquartile range $= Q_3 - Q_1$

$$= 37.0\,^\circ\text{C} - 36.68\,^\circ\text{C}$$

$$= 0.32\,^\circ\text{C}$$

This gives the spread of the temperatures for the middle 50% of the pupils in the sample.

Practical Follow-up **1.14**

1. Find the interquartile range for the data from your Group B practicals (Practical 1.3, 1.4 or 1.5).
2. Draw a box-and-whisker diagram for each set of data that you have.
3. If you have data for two groups ('A' sides and 'B' sides; males and females), how do they compare? What conclusions can you draw from your diagram?

Section **1.20**   **Variance and Standard Deviation**

A useful measure of variability would be one which would give a measure of the variation of observations relative to the mean. We could try to find the average deviation about the mean by summing all the values of $x_i - \bar{x}$ and dividing by n. Unfortunately the sum of $x_i - \bar{x}$ always equals zero (for a proof of this, see the end of this chapter). This is because positive and negative differences cancel each other out and so, to avoid this difficulty, they are squared (thus making them all positive).

The Variance

The variance is defined as

$$\text{Var}(x) = \frac{\sum_i (x_i - \bar{x})^2}{n}$$

for a set of observations $x_1, x_2, \ldots, x_n$ or

$$\text{Var}(x) = \frac{\sum_i f_i (x_i - \bar{x})^2}{\sum_i f_i}$$

for a frequency table. The variance gives the average of the square deviations about the mean (its square root is called the standard deviation).

As the mean of any set of figures is rarely an integer value, the formulae given above may be very difficult to use, e.g. if a value of $x_i - \bar{x}$ is given to two decimal places then $(x_i - \bar{x})^2$ will have four decimal places!

Alternative formulae are used and a proof that they are equivalent is given at the end of the chapter.

Alternative Formulae for Calculating the Variance

For a set of individual observations $x_1, x_2, \ldots, x_n$,

$$\text{Var}(x) = \frac{1}{n} \sum_i (x_i - \bar{x})^2$$

or

$$\text{Var}(x) = \frac{\sum_i x_i^2}{n} - (\bar{x})^2$$

For data given in a frequency table,

$$\mathrm{Var}(x) = \frac{\sum_i f_i(x_i - \bar{x})^2}{\sum f_i}$$

or

$$\mathrm{Var}(x) = \frac{\sum_i f_i x_i^2}{\sum_i f_i} - (\bar{x})^2$$

Calculating the Variance for Discrete Data

Here we calculate the variance of the number of working days taken for second-class letters to be delivered (Table 1.31).

Table 1.31   Number of working days taken for second-class letters to be delivered

Number x_i of days	f_i	$f_i x_i$	x_i^2	$f_i x_i^2$
1	8	8	1	8
2	55	110	4	220
3	31	93	9	279
4	8	32	16	128
5	0	0	25	0
6	0	0	36	0
7	0	0	49	0
8	1	8	64	64
9	0	0	81	0
10	1	10	100	100
		Total	261	799

$$\mathrm{Var}(x) = \frac{\sum_i f_i x_i^2}{\sum_i f_i} - \left(\frac{\sum_i f_i x_i}{\sum_i f_i} \right)^2$$

$$= \frac{799}{104} - \left(\frac{261}{104} \right)^2$$

$$= 7.683 - 6.298$$

$$= 1.385$$

Therefore,

standard deviation $= \sqrt{1.385} = 1.18$ (to two decimal places)

Strictly speaking, the x_i^2 column is not necessary, but it is a useful aid to working as $f_i x_i^2$ means f_i multiplied by x_i^2 and not $(f_i x_i)^2$. If omitted, use $f_i x_i^2 = f_i x_i(x_i)$.

Practical Follow-up **1.15**

1. Calculate the variance and standard deviation for the number of days taken for first-class letters from our data in Table 1.4.
2. How do these figures compare with those for second-class letters? Which set has the greater variance? Why is this?
3. Calculate the variances of the sets of data which you have collected for Practical 1.1 or 1.2. What conclusions can you draw?

Section **1.21** **Calculation of the Variance for Continuous Grouped Data**

Refer to the frequency table for the part-time earnings of male students given in Table 1.20.
The formula for variance is

$$\mathrm{Var}(x) = \frac{\sum_i f_i x_i^2}{\sum_i f_i} - \left(\frac{\sum_i f_i x_i}{\sum_i f_i} \right)^2$$

The middle of each group is used as x_i for calculating the variance, as it was when calculating the mean in Section 1.12. The final column $f_i x_i^2$ can involve some very large numbers unless the figures are simplified by a change of origin (and possibly a change of units also).

For our calculation, we have used an origin of £12.505 and units of £5.00. This has made our working relatively simple. Note that, as a measure of variability within the group, the variance is *not* affected by a change of origin (this would affect the mean, however), but, as units of £5.00 are used, the standard deviation must be multiplied by 5, and the variance must be multiplied by 25 to compensate.

The proof of this is given at the end of this chapter.

Variance and Standard Deviation of Male Sixth-formers' Earnings

Change the origin and units (Table 1.32).

Table 1.32 Earnings of male sixth-formers

Earnings (£)	Middle of group (£)	Origin d_i at £12.50 (units of £5)	f_i	$f_i d_i$	$f_i d_i^2$
0.01–5.00	2.505	−2	4	−8	16
5.01–10.00	7.505	−1	5	−5	5
10.01–15.00	12.505	0	6	0	0
15.01–20.00	17.505	1	9	9	9
20.01–25.00	22.505	2	4	8	16
25.01–30.00	27.505	3	1	3	9
30.01–35.00	32.505	4	1	4	16
		Total	30	−13 + 24	71

$$\sum_i f_i d_i = 11$$

$$\sum_i f_i d_i^2 = 71$$

origin $= £12.505$ (this only affects the mean)

units $= 5$

Now

$$\text{variance} = 5^2 \left[\frac{\sum_i f_i d_i}{\sum_i f_i} - \left(\frac{\sum_i f_i d_i}{\sum_i f_i} \right)^2 \right]$$

$$= 25 \left[\frac{71}{30} - \left(\frac{11}{30} \right)^2 \right]$$

$$= 25(2.36 - 0.134)$$

$$= 25 \times 2.232$$

$$= 55.806$$

Therefore

standard deviation $= £7.47$

Practical Follow-up 1.16

1. Calculate the variance and standard deviation of the earnings of female sixth formers using our figures in Table 1.27. Compare your answer with the variance for male sixth-formers' earnings.
2. Compare the variances of the two sets of data collected for your Group B practical study (Practical 1.3, 1.4 or 1.5).
3. Even with a calculator the use of a change of origin can be useful. Without it, you may be subtracting two very large and nearly equal numbers and so lose accuracy. For example, use your calculator to find the variance of 100 000 001, 100 000 002 and 100 000 003. Compare this with the variance of 1, 2 and 3. (Your answers should be equal!)

Section 1.22   Conclusions

Collect together your results, diagrams and follow-up work for each of your practical studies.

Group A Practicals

Practical 1.1

Did you study crosswords from *The Guardian* and the *Evening News* as we did? Were your results similar to ours?

We found that *The Guardian* crosswords had a higher mean for the lengths of words used, so that on average the words were longer by 1.68 letters. Also the words used in *The Guardian* crosswords were more variable in length than those in the *Evening News*. Do you consider that the measure we have used (i.e. length of words) gives us sufficient evidence to conclude that *The Guardian* crosswords are more difficult? What other measures of difficulty can you suggest?

Practical 1.2

We found that our samples of first- and second-class letters had very different distributions for the number of working days to delivery. The bulk of the first-class letters (80% in our sample) were delivered on the next working day. However, there were a few letters which took up to 10 working days (these tended to be larger packets which had been paid at the first-class rate). This had the effect of giving first-class letters a more skewed distribution than second-class letters. Deciding whether first-class or second-class post is better value is a difficult problem. However, we could calculate the average number of days for delivery divided by the cost: for first-class letters, this is $1.7/18 = 0.09$; for second-class letters, this is $2.5/13 = 0.19$.

This could be thought of as the mean number of days per penny cost, so that a lower value is better than a higher one. Using this measure, first-class post is twice as efficient as second-class post.

How do your results compare with ours? Are your conclusions different from ours?

Group B Practicals

Practical 1.3

In our study, we found that the average part-time earnings for male students was 92 p higher than for female students. The median wage for males, however, was over £2.00 higher. This shows that, while there are a few very low earners among the males (bringing down their mean value), the males at the centre of the group are earning higher wages than are their female counterparts. The variance for males was also greater.

The same proportions of male and female sixth-form students (about 1/3) do not have part-time jobs.

How do your results compare with ours? What conclusions can *you* draw from your results?

Practical 1.4

How do 'A' side and 'B' side tracks compare for pop singles? Were the 'A' side tracks longer? Were they less variable in length than the 'B' side tracks? Write up a report on the basis of your results.

Practical 1.5

Did you compare male and female temperatures? Which were higher on average (or was there little difference)? Were one group's temperatures more variable?

The Sum of Deviations About the Mean

To prove that

$$\sum_{i=1}^{n} (x_i - \bar{x}) = 0$$

we proceed as follows. The sum may be split and the two terms summed separately as

$$\sum_{i=1}^{n} (a_i + b_i) = \sum_{i=1}^{n} a_i + \sum_{i=1}^{n} b_i$$

Thus,

$$\sum_{i=1}^{n} (x_i - \bar{x}) = \sum_{i=1}^{n} x_i - \sum_{i=1}^{n} \bar{x}$$

For any set of observations, $\bar{x}$ is a constant and

$$\sum_{i=1}^{n} \bar{x} = n\bar{x}$$

since

$$\bar{x} = \frac{1}{n}\sum_i x_i$$

Therefore,

$$\sum_i (x_i - \bar{x}) = \sum_i x_i - n\bar{x}$$

However, the total for x_i $\left(\sum_{i=1}^{n} x_i\right)$ also equals $n\bar{x}$. Therefore,

$$\sum_{i=1}^{n} (x_i - \bar{x}) = n\bar{x} - n\bar{x}$$

$$= 0$$

Proof of the Alternative Formula for Variance

$$\text{Var}(x) = \frac{\sum_{i=1}^{n} (x_i - \bar{x})^2}{n}$$

$$= \frac{1}{n}\sum_{i=1}^{n} (x_i - \bar{x})^2$$

$$= \frac{1}{n}\sum_{i=1}^{n} (x_i^2 - 2x_i\bar{x} + \bar{x}^2)$$

$$= \frac{1}{n}\left(\sum_{i=1}^{n} x_i^2 - 2\bar{x}\sum_{i=1}^{n} x_i + \sum_{i=1}^{n} \bar{x}^2\right)$$

$$= \frac{1}{n}\left(\sum_{i=1}^{n} x_i^2 - 2\bar{x}n\bar{x} + n\bar{x}^2\right)$$

$$= \frac{1}{n}\left(\sum_{i=1}^{n} x_i^2 - n\bar{x}^2\right)$$

$$= \frac{\sum_{i=1}^{n} x_i^2}{n} - \frac{n\bar{x}^2}{n}$$

$$= \frac{\sum_{i=1}^{n} x_i^2}{n} - \bar{x}^2$$

Finding the Mean Using a Change of Origin and Units

If

working origin (or assumed mean) $= A$
units $ = U$ $\bigg\}$ (both constants)

So that

$$d_i = \frac{x_i - A}{U}$$

and

$$x_i = A + U d_i$$

then

$$\bar{x} = \frac{\sum\limits_i f_i x_i}{\sum\limits_i f_i}$$

$$= \frac{f_1 x_1 + f_2 x_2 + \ldots + f_n x_n}{\sum\limits_i f_i}$$

$$= \frac{f_1(A + U d_1) + f_2(A + U d_2) + \ldots + f_n(A + U d_n)}{\sum\limits_i f_i}$$

$$= \frac{A(f_1 + f_2 + \ldots + f_n) + U(f_1 d_1 + f_2 d_2 + \ldots + f_n d_n)}{\sum\limits_i f_i}$$

$$= \frac{A \sum\limits_i f_i}{\sum\limits_i f_i} + \frac{U \sum\limits_i f_i d_i}{\sum\limits_i f_i}$$

$$\bar{x} = A + U \bar{d}$$

Finding the Variance Using a Change of Origin and Units

If

working origin (or assumed mean) $= A$

units $\qquad\qquad\qquad = U$

$$d_i = \frac{x_i - A}{U}$$

and

$$x_i = A + U d_i$$

then

$$\mathrm{Var}(x) = \frac{\sum\limits_i f_i (x_i - \bar{x})^2}{\sum\limits_i f_i}$$

$$= \frac{\sum_i f_i [A + Ud_i - (\overline{A + Ud})]^2}{\sum_i f_i}$$

$$= \frac{\sum_i f_i [A + Ud_i - (A + U\bar{d})]^2}{\sum_i f_i} \qquad (A \text{ is a constant and } \bar{A} = A)$$

$$= \frac{\sum_i f_i (A + Ud_i - A - U\bar{d})^2}{\sum_i f_i}$$

$$= \frac{\sum_i f_i (Ud_i - U\bar{d})^2}{\sum_i f_i}$$

$$= \frac{U^2 \sum_i f_i (d_i - \bar{d})^2}{\sum_i f_i}$$

$$\text{Var}(x) = U^2 \, \text{Var}(d)$$

Chapter 2 Probability

Section 2.1 Introductory Practicals

Instructions are given here for each of the introductory practicals in this chapter, together with a full list of all the materials required. Recording sheets are included among the photocopiable pages (pages PC1–PC7).

As you work through each practical, enter your results on to the appropriate recording sheet so that everyone's results can be collected together. Subsequent sections in this chapter will analyse and discuss the practical results; so keep them carefully.

Practical 2.1   Heads or Tails?

Equipment

You require the following.

A coin.
Recording sheet 1 (page PC1).
A calculator.

Method

1. Toss the coin.
2. Record the result of this 'trial' as either H (heads) or T (tails) on the recording sheet.
3. Repeat this 20 times in all.
4. At the end of your 20 trials, count the number of heads.
5. Add your total to the previous running total for heads.
6. Calculate the proportion of heads for *all* the trials so far (using the calculator).

Thus,

$$\text{proportion of heads (so far)} = \frac{\text{running total for heads}}{\text{number of trials so far}}$$

Practical 2.2a   Throwing Pins I

Equipment

You require the following.

Recording sheet 2 (page PC2).
Sheet of paper with lines drawn 2.5 cm apart.
10 dressmaker's pins (2.5 cm long).

Method

1. Place the lined sheet on the table.
2. Throw the 10 pins at random onto the sheet.
3. Count and record the number of pins which land crossing a line.
4. Repeat the experiment 5 times.

The column labelled 'Running total for number of pins crossing lines' on the recording sheet will help you to calculate the proportion of all pins which landed on a line.

Practical 2.2b   Throwing Pins II

Equipment

You require the following.

Recording sheet 2 (page PC2).
Sheet of paper with lines drawn 5 cm apart.
10 dressmaker's pins (2.5 cm long).

Method
1. Place the lined sheet on the table.
2. Throw the 10 pins at random onto the sheet.
3. Count and record the number of pins landing across a line.
4. Repeat the experiment 5 times.
5. Calculate the cumulative probability that a pin lands across a line at the end of your 5 trials.

Practical 2.3   Throwing a Die

Equipment

You require the following.

A die.
Recording sheet 3 (page PC3).

Method

1. Throw the die until you score a 6.
2. Count the number of throws it took until you scored the 6.
3. Record that number on the left-hand side of the recording sheet.
4. Repeat this for 5 trials.
5. Enter your 5 results in the tally chart on the right-hand half of the recording sheet.

Practical 2.4   Pushing Pennies

Equipment

You require the following.

A Penny.
A 'push penny' grid (page PC4). The sheet that we provide is very small and will produce 'boundary effects'. For that reason, you are advised to draw a larger one (still with 4 cm squares).
Recording sheet 4 (page PC5).

Method

1. Place the penny anywhere on the white margin. Push the penny with a ruler so that it slides across onto the grid. If the penny stops outside the grid, try again.
2. If your penny stops completely inside a square, you *win*. If it lands across a line, you *lose*.
3. Record your results as W (win) or L (lose).
4. Repeat the experiment 10 times.
5. At the end of your 10 trials, count the number of wins, and add this on to a running total of wins.
6. Calculate the proportion of wins cumulatively after every 10 trials.

Practical **2.5   Shaking Cards in a Bag**

Equipment

You require the following.

3 cards.
A cloth bag.
Recording sheet 5 (page PC6).

The cards should be as in Figure 2.1.

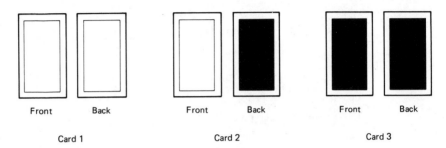

Figure 2.1   The three cards required for Practical 2.5

Method

1. Shake the 3 cards in the bag.
2. Pick a card.
3. *Without looking at the under side*, place the card on the table. You may look at the top side.
4. What colour is the top side? (Given that the top side is white, say, your card must be card 1 or card 2 (it cannot be card 3). Proportionally, how often do you think the under side will be white? Half the time perhaps?)
5. Look at the under side.
6. Record the under-side colour as B (black) or W (white) on the correct grid on the sheet.
7. Replace your card and shake the bag.
8. Repeat the experiment 10 times in all.

Practical **2.6a   Throwing Dice I**

Equipment

You require the following.

Die A and Die B.
Recording sheet 6 (page PC7).

Method

1. Throw the 2 dice provided. The die showing the higher score wins.
2. Record the two scores and the winner.
3. Repeat this 5 times (Figure 2.2).

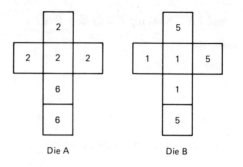

Figure 2.2 Practical 2.6a

Practical 2.6b   Throwing Dice II

Equipment

You require the following.

Die B and Die C.
Recording sheet 6 (page PC7).

Method

1. Throw the 2 dice provided. The die showing the higher score wins.
2. Record the two scores and the winner.
3. Repeat this 5 times (Figure 2.3).

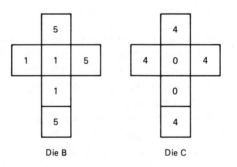

Figure 2.3 Practical 2.6b

Practical 2.6c   Throwing Dice III

Equipment

You require the following.

Die C and Die D.
Recording sheet 6 (page PC7).

Method

1. Throw the 2 dice provided. The die showing the higher score wins.
2. Record the two scores and the winner.
3. Repeat this 5 times (Figure 2.4).

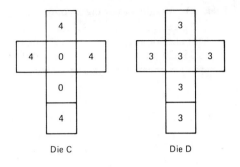

Figure 2.4   Practical 2.6c

Practical 2.6d   **Throwing Dice IV**

Equipment

You require the following.

Die A and Die D.
Recording sheet 6 (page PC7).

Method

1. Throw the 2 dice provided. The die showing the higher score wins.
2. Record the two scores and the winner.
3. Repeat this 5 times (Figure 2.5).

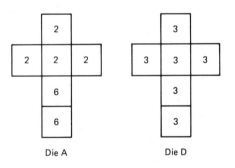

Figure 2.5   Practical 2.6d

Section 2.2   **Results of the Experiments**

Practical 2.1

Table 2.1 shows our results for 20 repeats each of 20 tosses of a coin (i.e. 400 tosses in all).

We have plotted the cumulative proportion of heads, calculated after each 20 trials on the graph in Figure 2.6.

These results are unusual, in that our first three groups of 20 all gave the same proportion, 0.55. After that, there are wide fluctuations but at 300 trials the proportion of heads appears to be settling down to a value between 0.49 and 0.5. Figure 2.7 is the graph drawn for another set of results for 20 groups of 100 trials. Again there are wide fluctuations in the early results, but at 1300 trials the proportion of heads is 0.4992 and from then onwards

Table 2.1 Results of tossing a coin

Number of heads in 20 trials	Cumulative total of heads	Cumulative proportion of heads
11	11	$\dfrac{11}{20} = 0.55$
11	22	$\dfrac{22}{40} = 0.55$
11	33	$\dfrac{33}{60} = 0.55$
8	41	$\dfrac{41}{80} = 0.5125$
9	50	$\dfrac{50}{100} = 0.5$
8	58	$\dfrac{58}{120} \approx 0.4833$
7	65	$\dfrac{65}{140} \approx 0.4643$
10	75	$\dfrac{75}{160} \approx 0.4688$
12	87	$\dfrac{87}{180} \approx 0.4833$
9	96	$\dfrac{96}{200} = 0.48$
14	110	$\dfrac{110}{220} = 0.5$
9	119	$\dfrac{119}{240} \approx 0.4958$
8	127	$\dfrac{127}{260} \approx 0.4885$
8	135	$\dfrac{135}{280} \approx 0.4821$
10	145	$\dfrac{145}{300} \approx 0.4833$
14	159	$\dfrac{159}{320} \approx 0.4969$
9	168	$\dfrac{168}{340} \approx 0.4941$
9	177	$\dfrac{177}{360} \approx 0.4917$
11	188	$\dfrac{188}{380} \approx 0.4947$
10	198	$\dfrac{198}{400} = 0.495$

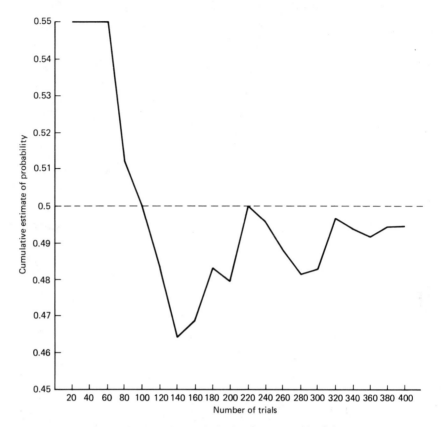

Figure 2.6 Proportion of heads plotted cumulatively after every 20 trials

the fluctuations are very slight around the value of 0.5. As the number of trials increases, on the whole these fluctuations will become smaller and smaller.

Here we have obtained experimentally the proportion of heads or the relative frequency of heads. From 400 trials, we obtained a relative frequency of 0.495 for heads. A second experiment of 2000 trials gave a relative frequency of 0.501. Both experiments have given an experimental or empirical estimate of the probability of obtaining a head with that coin.

By obtaining a large number of experimental results, we see that the relative frequency seems to approach a limiting value which is the probability. The empirical probability is calculated from experimental results thus: •

$$\text{empirical probability for heads} = \frac{\text{number of trials in which heads occurs}}{\text{total number of trials}}$$

The larger the number of trials, the more likely is this empirical probability to be close to the true probability. For an unbiased coin which is equally likely to fall heads or tails, we would expect the long-term proportion of heads to be 1/2. This value is known as the theoretical probability of obtaining heads:

$$\text{theoretical probability for heads} = \frac{\text{number of ways in which a head can occur}}{\text{number of different equally likely outcomes}}$$

'Heads' is one outcome out of the two equally possible outcomes; so the theoretical probability of a head is 1/2. Note that this type of probability can only be worked out when the different outcomes are all equally likely.

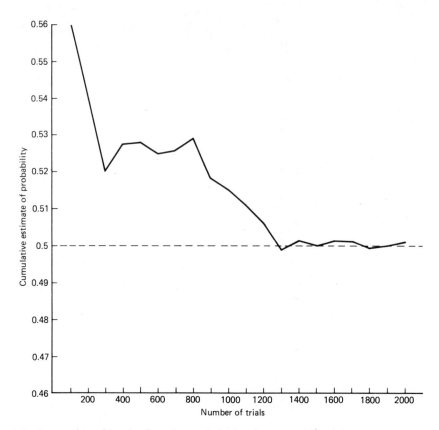

Figure 2.7   Proportion of heads plotted cumulatively after every 100 trials

In many situations, we cannot calculate a value for the theoretical probability. In those cases, we must use the empirical probability from large numbers of results. For instance the probability that a person in the population is affected by cystic fibrosis is given as 1 in 2000 (or 0.0005). This is an example of an empirical probability.

Practical Follow-up **2.1**

Practical 2.1

From your results for Practical 2.1, plot the cumulative probabilities for 'heads' on a graph similar to ours.

Practical 2.2a

Table 2.2 shows our results for 40 trials of Practical 2.2a. In each trial, 10 pins were dropped onto a lined sheet. The lines were 2.5 cm apart and the pins were 2.5 cm long. For each

Table 2.2   Number of pins falling across a line

7	7	6	5	6	4	5	7	8	7
7	4	7	8	5	6	6	5	7	6
5	4	7	5	7	6	7	5	8	8
6	3	4	7	5	8	7	7	4	6

trial, we recorded the number of pins which fell across a line. This experiment is often known as Buffon's needle (named after the Comte de Buffon who did this experiment a large number of times). In all, 242 pins, out of the 400 thrown, landed across lines. The empirical probability that a pin lands across a line is $242/400 = 0.605$.

Practical Follow-up 2.2

Calculate the empirical probability that a pin lands across a line from the results of your entire class.

Empirical Estimation of π

Theoretically the probability that a pin lands across a line is $2/\pi$. A proof of this is given at the end of this chapter.

Given that result, we can use our experimental data to estimate π, if we take the empirical probability as an experimental estimate of the theoretical probability:

empirical probability $= 0.605$

theoretical probability $= \dfrac{2}{\pi}$

If the empirical probability is approximately equal to the theoretical probability,

$$0.605 \approx \frac{2}{\pi}$$

$$\pi \approx \frac{2}{0.605}$$

$$\pi \approx 3.305$$

The estimate 3.305 which we have obtained from our data slightly over-estimates π, but it is based on the results of only 400 experimental trials.

Calculate an estimate for π based on your experimental results. Is your estimate closer than ours?

Practical 2.2b

From your experimental results, calculate the empirical probability that a pin 2.5 cm long lands across a line (where lines are 5 cm apart). How does this result compare with your result for Practical 2.2a (lines 2.5 cm apart)?

What do you think is the theoretical probability that a pin lands across the line? Why is this?

Has your experiment given you a good estimate for π?

Practical 2.3

Table 2.3 shows our results for the number of throws required to score 6, shown as a frequency table. We repeated the experiment 122 times in all.

The bulk of our results occur up to 16 throws, but there were 5 occasions on which more than 16 throws were required (including one trial in which 36 throws were needed!). From the frequency table, we calculated the mean number of throws required to be 6.475.

Table 2.3   Number of throws required to score a 6

Number x of throws	Frequency f	Number x of throws	Frequency f	Number x of throws	Frequency f
1	12	13	3	25	0
2	15	14	1	26	0
3	17	15	2	27	0
4	12	16	2	28	1
5	11	17	0	29	0
6	6	18	0	30	0
7	12	19	0	31	0
8	6	20	1	32	0
9	7	21	0	33	0
10	6	22	1	34	0
11	2	23	0	35	0
12	3	24	1	36	1

Practical Follow-up **2.3**

Draw up a frequency table for the number of throws required to score a 6, and calculate the mean number of throws. How does your result compare with ours?

Practical Follow-up **2.4**

Practical 2.4

Look at the class results for Practical 2.4 and find the totals for the following.

1. Wins.
2. Losses.
3. Total number of games played.

From these, calculate the empirical (experimental) probabilities for a win and a loss.

If you have calculated the empirical probability of a win cumulatively after every 20 trials, you may plot your results graphically as we did for Practical 2.1.

The Theoretical Probabilities

1. Measure a square on the push penny grid (page PC4) and find its area.
2. Measure the diameter of a penny and find its radius.

To win at push penny, the coin must lie completely inside a grid square and cannot cross any lines. The centre of the penny must land within a particular area inside the grid square as in Figure 2.8.

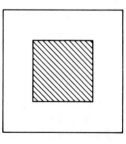

Figure 2.8   The centre of the coin must land within the shaded area

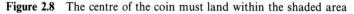

Can you decide on the dimensions of the shaded area? What is the theoretical probability of a win? In this analysis, we have ignored any 'boundary effects' which will be most pronounced on a small grid. For this reason you are advised to draw your own grid on a larger sheet rather than the grid provided.

Practical Follow-up **2.5**

Practical 2.5

Find the class results for this experiment. Firstly, concentrate on the totals for those choices in which the top side of the card was white.

1. What is the empirical probability that the under side is white?
2. What is the empirical probability that the under side is black?

Are these results as you expected? As the top side of the card is white, we know that we must have white/white or white/black. At first, these two alternatives appear equally likely, but are they?

Theoretical Probabilities

Figure 2.9 is a diagram showing all the possible outcomes for this experiment. Each of the three cards is equally likely to be chosen from the bag. Whichever card is chosen, either of the two sides could appear on top. Altogether there are six sides (three black and three white) any of which could appear as the top side.

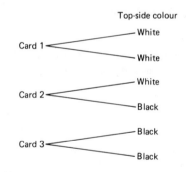

Figure 2.9 Colour of the top side

Looking at your experimental results, you should find that roughly half of all your trials had white as the top side.

Next we can consider the under side (Figure 2.10). Of the six outcomes possible, three

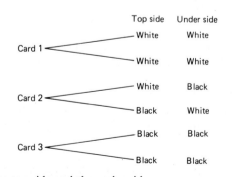

Figure 2.10 Colours of the top side and the under side

have white as the top-side colour but, of these three, only *one* is white/black but *two* are white/white (either side of the white/white card may appear first). Thus, your experimental results should yield roughly half as many with black on the under side as white.

Similarly, your results for top side black should give approximately 2/3 as black/black and 1/3 as black/white.

Practical Follow-up 2.6

Practical 2.6

1. From your results find the winners for each pair of dice (Figure 2.11). Do you find any of these results surprising?
2. For each pair find the proportion of the games won by the overall winner.

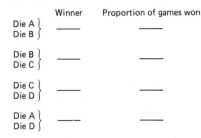

Figure 2.11   The winners for each pair of dice

Theoretical Probabilities

Die A and Die B

We can list the different outcomes for both dice.
For Die A, the outcomes are 2 or 6:

$$P(A = 2) = \frac{4}{6} = \frac{2}{3}$$

$$P(A = 6) = \frac{2}{6} = \frac{1}{3}$$

For die B, the outcomes are 1 or 5:

$$P(B = 1) = \frac{3}{6} = \frac{1}{2}$$

$$P(B = 5) = \frac{3}{6} = \frac{1}{2}$$

There are various types of diagram which can be drawn to represent all the outcomes possible when both dice are thrown together. You may already be familiar with the type of diagram known as a *tree diagram*. This diagram is built up in stages to show multiple events. First, we show the two outcomes for die A (with their probabilities) as the first two branches of the 'tree' (Figure 2.12). Next we show the outcomes for die B to give all the 4 combinations possible. These are as follows:

(2, 1) (2, 5) (6, 1) (6, 5)

Figure 2.13 is the finished tree diagram.

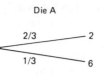

Figure 2.12   The two outcomes for die A

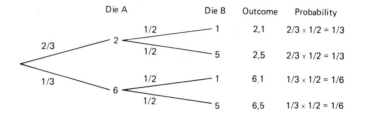

Figure 2.13   The finished tree diagram

Note that the total of all the probabilities for the 4 possible outcomes equals 1. To calculate the probability for a combined event, e.g. a 2 on die A *and* a 1 on die B, the probabilities on the branches of the tree must be multiplied together.

You can look at this in the following way.

On 2/3 of the occasions, die A will show a 2 and, on 1/2 of *these* occasions, die B will then show a 1. Hence, on 1/3 of all occasions, a 2 on A will be followed by a 1 on B. The rule is

probability $(A = 2$ *and* $B = 1)$ = probability $(A = 2)$ × probability $(B = 1$ given that $A = 2)$

where A indicates the score on die A and B indicates the score on die B; this is an example of the general result

$$P(A \text{ and } B) = P(A) \times P(B \text{ given } A)$$

$P(B$ given $A)$ is called the conditional probability of B given A and is written $P(B|A)$. Also, $P(A$ and $B)$ is often written as $P(A \cap B)$ so that

$$P(A \cap B) = P(A)P(B|A)$$

In this case, the probability that B takes a particular value such as 1 does not depend on the value of A; so the probability of B given A is the same as the probability of B.

Events A and B are independent and

$$P(A \cap B) = P(A)P(B)$$

This is known as the *multiplication law*.

To calculate the probability that die A wins, we must find those outcomes for which A gets a higher score than B. These outcomes are

2, 1 probability $= 1/3$

or

6, 1 probability $= 1/6$

or

6, 5 probability $= 1/6$

Thus the event (2, 1) occurs on 1/3 of all occasions, (6, 1) occurs on 1/6 of all occasions and (6, 5) on 1/6 of all occasions. The total proportion of all occasions on which *any one* of these events occurs is $1/3 + 1/6 + 1/6 = 2/3$.

In situations in which any one of several outcomes can be considered as alternatives, their probabilities may be added to find the total probability that *any one* of them occurs (provided that these outcomes are mutually exclusive, i.e. they cannot occur at the same time).

The formal statement of this idea is known as the *addition law*.

For any two events *A* and *B*, the probability that either A or B or both occurs is written as $P(A \cup B)$. Therefore,

$$P(A \cup B) = P(A) + P(B) - P(A \cap B)$$

However, if *A* and *B* are mutually exclusive and $P(A \cap B) = 0$, we have a simpler statement

$$P(A \cup B) = P(A) + P(B) \text{ for mutually exclusive events}$$

Now compare your experimental results with the theoretical probability that die A has a higher score than die B. Did you find that die A was the winner for about 2/3 of your trials?

Draw tree diagrams for the other 3 pairs of dice, and compare the theoretical probabilities for the winner of each pair with your experimental results.

Section 2.3 **Alternative Probability Experiments**

Practical 2.7 **Probability that Two Students Have the Same Birthday**

By using school registers (or class lists giving dates of birth), find out how many classes have at least two people with the same birthday (for classes of 26 people the empirical probability that at least two have the same birthday is about 1/2).

Practical 2.8 **Dominoes**

Place a set of dominoes in a large bag and pick out two at random. Do they 'match', i.e. could they be placed next to each other in a game? Repeat the experiment 100 times and find the empirical probability of drawing two dominoes which 'match'.

Find the theoretical probability of a match. Are your experimental results in line with this?

Practical 2.9 **People Sitting in a Student Cafeteria**

This is a longer study which you can try for yourself. Alternatively, you will be able to decide whether the findings reported here are in line with the theoretical probabilities.

The survey was originally conducted by a psychologist who observed people sitting in a student cafeteria, where the tables were all square in shape, and seated 4 people only.

As people came in, he looked especially for people arriving in pairs or those arriving alone but meeting a friend already in the cafeteria. If the two people sat together on an otherwise empty table, he noted their seating pattern. In addition, he looked at the seating arrangements for pairs of people who were not obviously friendly or interacting but who just happened to be sitting at the same table.

He concentrated then on those tables at which two people only were seated. He noted whether they were sitting in opposite seats or in adjacent seats on any corner. Table 2.4 shows his findings. Pairs who were interacting (talking or working together for instance) seem to prefer adjacent corner seats, or do they? Are these results in line with those which we would expect purely by chance?

Table 2.4 Two people sitting at a table for four

	Friendly (or interacting)	Not interacting
Opposite seats	63	16
Adjacent seats	134	2

Theoretical Probabilities

Draw all the different possible seating patterns for two people sitting at a table arranged as in Figure 2.14.

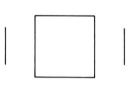

Figure 2.14 Arrangement for a square table seating four people

How many of these alternatives involve the two people in sitting at opposite seats? How many are in adjacent seats?

What is the theoretical probability that two people sitting *at random* will have chosen adjacent seats?

Has the psychologist proved that friends prefer to sit in adjacent seats?

If you feel that he may have proved that non-interacting pairs have a preference for opposite seats, you will be able to test this hypothesis later in the course (in Chapter 13).

Section 2.4 Genetic Inheritance and Probability

Albinos are people born with no pigment (melanin) in their skin, hair or iris of the eye. As a result, their skin is very pale, their hair is white and their irises are pink because the eye colouring pigment, normally present, is not there to mask the blood vessels in the eye. For a child to be born an albino, it must inherit the albino gene from *both* parents. This gene is very rare, and only one person in 70 is a carrier. Carriers have the albino gene (which is recessive) and a normal gene which is dominant. The dominant normal gene suppresses the albino gene so that the carrier is quite normal in appearance.

On the assumption of independence, the chance that two carriers will marry is $(1/70) \times (1/70) = 1/4900$. Figure 2.15 is a tree diagram to show the possible marriages of carriers and non-carriers.

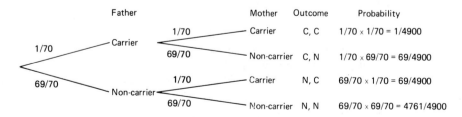

Figure 2.15 Possible marriages of carriers (C) and non-carriers (N) of the albino gene

Of all the marriages possible between carriers and non-carriers, it is only the first category (two carriers) which can give rise to albino children. Even so, not all the children in such a marriage will be affected. Each parent has a normal gene in addition to the albino gene; so we can now consider the probability that both parents will pass on the recessive albino gene to any child (Figure 2.16).

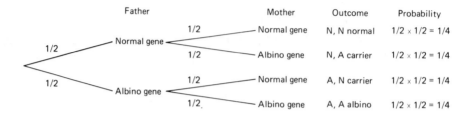

Figure 2.16 Genetic inheritance of the albino gene

Thus, each child born in such a marriage has a probability of $1/2$ of being a carrier of albinism, $1/4$ of being completely normal and only $1/4$ of being an albino.

To calculate the proportion of albinos in the population at large, we need to consider two outcomes which must occur together.

1. Both parents must be carriers $(p = (1/70) \times (1/70) = 1/4900)$.
and
2. The child inherits two albino genes $(p = 1/4)$.

Thus the probability that a person is an albino is approximately $(1/4900) \times (1/4) = 1/19\,600$ or roughly 1 in 20 000.

Even Rarer Cases

If two albinos marry, they will certainly produce an albino child. The probability that this marriage happens at random would be

$$\frac{1}{19\,600} \times \frac{1}{19\,600} = \frac{1}{384\,160\,000}$$

approximately 1 chance in 400 million. (You could argue that this cannot be considered as a random event!)

However, an albino might marry a carrier. This would happen with a probability of

$$\frac{1}{19\,600} \times \frac{1}{70} = \frac{1}{1\,372\,000}$$

(less than 1 chance in a million!).

Draw a tree diagram to show how they may pass on the relevant genes to their children. What is the probability that any child born in that marriage will be albino?

Thirdly, we can consider the case in which an albino marries a person who is not a carrier.

Draw a tree diagram to show the combinations of genes which they may pass on to their children. What is the probability that they produce an albino child?

Practical Follow-up 2.7

Find the probabilities that the following marriages produce *carriers* of albinism.

1. Carrier, normal.
2. Carrier, carrier.
3. Carrier, albino.
4. Albino, normal.

Other inherited conditions caused by recessive genes include sickle cell anaemia and blue eyes (brown eyes are dominant!).

Section 2.5   What Proportion of Babies are Boys?

In Britain, where there is no large-scale medical intervention, the ratio of male to female births remains fairly constant at about 105 boys to every 100 girls. In societies with poorer standards of medical care, this excess of boys disappears during childhood, as boys have a higher infant mortality rate and are more likely to succumb to disease. Cultural factors may also intervene; if boys are encouraged to play more adventurous games (or are given more dangerous work), they are more likely to be involved in accidents. This was certainly true in Britain 100 years ago, when the sex ratio at the age of marriage was only 90 men to 100 women. Today, our improved health care keeps more boys alive and there is now a slight excess of males in the 20–30 year age group.

Practical Suggestion

As a longer-term project over several weeks, record the births reported in your local paper. Calculate the ratio of boys to girls and see whether it is close to 105 to 100. Be careful, however, as you might find that the birth of a boy may be more likely to be reported than the birth of a girl—particularly in a paper such as *The Times*.

If parents are able to choose the sex of their child, will this naturally occurring ratio of boys to girls change significantly? To date, there has not been any major survey of the preferences of newly married couples, regarding the size and composition of their 'ideal' family.

While it is difficult and expensive to pre-determine the sex of a child at conception, there are methods of determining a child's sex later on in pregnancy. On 28 September 1986, *The Observer* reported a birth trend which is worrying health officials in Korea. Apparently the ratio of male to female births is now in the region of 117 to 100.

The health service in Korea is mainly private, but ultrasound scanning is available very cheaply, for the purpose of detecting physical abnormalities. It is frequently possible during a scan for a doctor to see whether the baby is a boy or a girl. The Korean government have warned doctors not to reveal this information to parents, but clearly this warning is not being heeded. Apparently, many more abortions now involve girls rather than boys.

Korean culture dictates that a girl will leave home to live with her husband's family while the son takes responsibility for looking after his ageing parents (there are no state pensions). Ancestral ties are important too and many families can trace their family tree back for 2000 years through the male line. For these reasons, parents in Korea wish to ensure that they have at least one son.

To calculate the effects of this preference, several factors must be taken into account. Firstly, will parents try to determine the sex of their first baby, or are they more likely to leave the sex of that baby to chance and try to exercise choice at later births? Secondly the size of the family is important too and sex preferences exert a lesser effect in societies with larger families (than in those where the average family is only two children).

As our first example, we shall show what happens in two-child families, where parents could and do use sex selection to ensure that their second child is a boy, if their first child is a girl. (If the first child is a boy, we assume that they will be willing to leave the sex of the second child to chance, as they did for the first child.)

Suppose this were to happen in Britain. Then, out of every 205 children born naturally, 105 are boys. Thus the probability of a boy is 0.51 where there is no intervention (Figure 2.17).

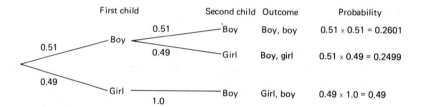

Figure 2.17 The probability of a boy or girl in a two-child family (when girl, girl possibility is excluded)

To calculate the sex ratio of births in this situation, we can multiply each probability by 100 (to give expected frequencies for 100 births) (Table 2.5). For example, out of 100 children in two-child families, 26 will be in boy, boy families; so we have 26 boys and no girls.

Table 2.5 Sex ratio of births for two-child families (when girl, girl possibility is excluded)

Outcome	Probability	Expected frequency	Boys	Girls
Boy, boy	0.2601	26	26	0
Boy, girl	0.2499	25	12.5	12.5
Girl, boy	0.49	49	24.5	24.5
		Total	63.0	37.0

There should be 25 one-boy one-girl families, giving us 12.5 boys and 12.5 girls(!) and so on.

The ratio of boys to girls at the end of these calculations is 63 to 37. We can recalculate this, as the ratio of boys to every 100 girls:

boys to girls

63 to 37

$$\frac{63}{37} \times 100 \text{ to } 100$$

170 to 100

Thus, if all families consisted of two children only and if parents used sex selection on the second child to ensure at least one boy, the ratio of boys to girls would be 170 to 100.

Figure 2.18 shows some similar calculations for three-child families. Here the effect on the sex ratio is reduced, as parents only use selection on the third child to ensure that a boy is born (if they have two girls to date).

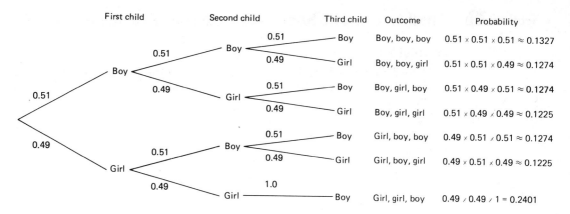

Figure 2.18 The probability of a boy or girl in a three-child family (when girl, girl, girl possibility is excluded)

To calculate the sex ratio, we can multiply each probability by 100 again, to give expected frequencies for 100 births (Table 2.6).

Table 2.6 Sex ratio of births for three-child families (when girl, girl, girl possibility is excluded)

Outcome	Probability	Expected frequency	Boys	Girls
Boy, boy, boy	0.1327	13.27	13.27	0
Boy, boy, girl	0.1274	12.74	8.48	4.24
Boy, girl, boy	0.1274	12.74	8.48	4.24
Boy, girl, girl	0.1225	12.25	4.08	8.16
Girl, boy, boy	0.1274	12.74	8.48	4.24
Girl, boy, girl	0.1225	12.25	4.08	8.16
Girl, girl, boy	0.2401	24.01	8.00	16.00
		Total	54.87	45.04

The ratio is approximately 55 boys to 45 girls.
We can recalculate this as the ratio of boys to every 100 girls:

boys to girls

55 to 45

$$\frac{55}{45} \times 100 \text{ to } 100$$

122 to 100

Coincidentally, this answer is quite close to the ratio now found in Korea. Of course, not all families will consist of three children (some families will be much larger) and of course not all parents will use sex determination. However, these two factors will tend to reduce the sex ratio to below 122 to 100 but are more difficult to take into account in a simple model such as ours.

Section **2.6** **Another Look at Some of the Earlier Practicals**

Practical 2.3

Find your results for Practical 2.3, and the work for the practical follow-up. In this section, we shall look at this experiment more theoretically.

Theoretical Probabilities for Practical 2.3

At which throw, is a 6 most likely to appear? If we look at our experimental results, the mode of our frequency distribution is 3; so our data would suggest this as the answer (Figure 2.19).

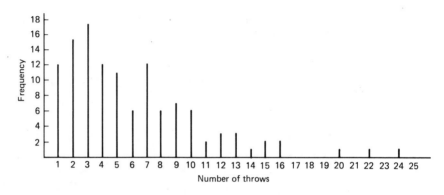

Figure 2.19   The number of throws required to score a 6 on a die

Find the mode of your frequency distribution. Almost certainly, your answer is different from ours (we do not have a sufficiently large number of trials for a consistent pattern to have emerged).

We shall try to calculate theoretical probabilities for X, the number of throws, in order to find the value of X with the highest probability. In this way, we should be able to decide when we are most likely to throw a 6. (We use X to stand for the number of throws in general and x for a particular value. So we have $P(X = x)$ meaning the probability that we get a 6 in exactly x throws. This will also be shortened to be written $p(x)$.)

Here are the theoretical probabilities for throwing a 6 in each throw.

First Throw

The probability that we score a 6 on the first throw is 1/6.

$$P(X = 1) = p(1) = \frac{1}{6}$$

Second Throw

The probability that we score a 6 on the second throw has two components.

1. We must have failed to get a 6 on the first throw; $p = 5/6$.
2. We must score a 6 on the second throw: $p = 1/6$.

The probability that these two events occur together is

$$P(X = 2) = p(2) = \frac{5}{6} \times \frac{1}{6}$$

$$= \frac{5}{36}$$

Third Throw

For our first 6 to occur at the third throw, all three of the following must have occurred.

1. We must have failed to get a 6 on the first throw.
2. We must have failed to get a 6 on the second throw.
3. We must have succeeded to throw a 6 on the third throw.

Thus,

$$P(X = 3) = p(3) = \frac{5}{6} \times \frac{5}{6} \times \frac{1}{6}$$

or

$$p(3) = \frac{1}{6} \times \left(\frac{5}{6}\right)^2$$

The theoretical probabilities for 4, 5, 6, etc., throws are calculated in the same way (Table 2.7).

The highest probability occurs at $X = 1$, i.e. we are more likely to score our *first 6* on the first throw than any other. The probability that we score our *first 6* on subsequent throws is reduced by a factor of 5/6 every time. Theoretically, then the mode of the distribution is 1.

Note that the probabilities form a geometric series with a first term 1/6 and each subsequent term multiplied by 5/6 (the common ratio):

$$a = \frac{1}{6} \text{ (first term)}$$

$$r = \frac{5}{6} \text{ (common ratio)}$$

The nth term of the series is given by

$$\frac{1}{6} \times \left(\frac{5}{6}\right)^{n-1}$$

A distribution of probabilities forming a series such as this is called a *geometric distribution*.

We can compare our experimental results with the theoretical probabilities if we calculate the frequencies that we would expect from those probabilities. For instance, to calculate the expected frequencies for our experiment of 122 trials, simply multiply each probability by 122.

Our expected frequencies have been given to two decimal places, on the table, even though in fact frequencies must be integers (and decimal fractions do not make sense). We have shown the experimental frequencies and the expected frequencies on the same graph in Figure 2.20 (expected values do not have to be integers even for integer variables, see the discussion of χ^2 in Chapter 13).

Table 2.7 Expected frequency of throwing a 6 (for 122 trials)

Number of throws needed to score a six	Theoretical probability $p(x)$	Expected frequency $122p(x)$
1	$\dfrac{1}{6} \approx 0.167$	20.33
2	$\dfrac{1}{6} \times \dfrac{5}{6} \approx 0.139$	16.94
3	$\dfrac{1}{6} \times \left(\dfrac{5}{6}\right)^2 \approx 0.116$	14.12
4	$\dfrac{1}{6} \times \left(\dfrac{5}{6}\right)^3 \approx 0.096$	11.77
5	$\dfrac{1}{6} \times \left(\dfrac{5}{6}\right)^4 \approx 0.080$	9.81
6	$\dfrac{1}{6} \times \left(\dfrac{5}{6}\right)^5 \approx 0.067$	8.17
7	$\dfrac{1}{6} \times \left(\dfrac{5}{6}\right)^6 \approx 0.056$	6.80
8	$\dfrac{1}{6} \times \left(\dfrac{5}{6}\right)^7 \approx 0.047$	5.67
9	$\dfrac{1}{6} \times \left(\dfrac{5}{6}\right)^8 \approx 0.039$	4.73
10	$\dfrac{1}{6} \times \left(\dfrac{5}{6}\right)^9 \approx 0.032$	3.94
11	$\dfrac{1}{6} \times \left(\dfrac{5}{6}\right)^{10} \approx 0.027$	3.28
12	$\dfrac{1}{6} \times \left(\dfrac{5}{6}\right)^{11} \approx 0.022$	2.73
13	$\dfrac{1}{6} \times \left(\dfrac{5}{6}\right)^{12} \approx 0.019$	2.28
14	$\dfrac{1}{6} \times \left(\dfrac{5}{6}\right)^{13} \approx 0.016$	1.90

We have not included expected frequencies for values of x over 21 as the expected frequencies were zero to the nearest integer. However, a tiny probability does exist for values of $x > 21$. For example,

$$P(X = 22) = \left(\dfrac{5}{6}\right)^{21} \times \dfrac{1}{6} \approx 0.00362$$

Although the probability gets smaller as the number of throws increases, the probability will never equal exactly zero. The probabilities form an infinite series. You can check that the sum of the probabilities equals 1 but you need to know that the sum of an infinite geometric series is given by $a/(1-r)$, where a is the first term equal to 1/6 and r is the common equal to 5/6. So the total of all the terms is $(1/6)/(1 - 5/6) = (1/6)/(1/6) = 1$.

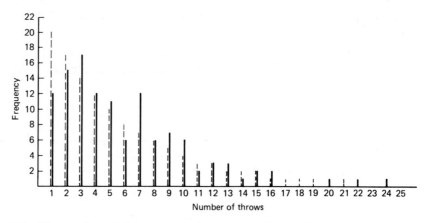

Figure 2.20 The number of throws required to score a 6 on a die: ——, experimental results; ---, theoretical expected frequencies

Practical Follow-up 2.8

Calculate the expected frequencies for your experiment. Your answers will depend on the total number of trials in your experiment as

expected frequency = theoretical probability × total frequency

Draw a line graph to show these expected frequencies and your experimental results.

Theoretical Value for the Mean Number of Throws Required to Score a 6

The mean of a frequency distribution is obtained from

$$\bar{x} = \frac{\sum_i f_i x_i}{\sum_i f_i}$$

We used this formula to calculate the mean from our experimental results, but we would like to calculate the theoretical value for the mean also.

We know that the sample mean

$$\bar{x} = \frac{\sum_i f_i x_i}{\sum_i f_i}$$

which can be rewritten as

$$\bar{x} = \sum_i x_i \frac{f_i}{\sum_i f_i}$$

Now $f_i / \sum_i f_i$ is the relative frequency of getting the value of x_i. So, if we take larger and larger samples, this becomes $p(x_i)$, the probability that $X = x_i$. Therefore, the mean of the theoretical distribution (which we denote by μ to distinguish it from the sample mean) is

$$\mu = \sum_i x_i p(x_i)$$

This is much easier to work with and we can calculate the theoretical value of the mean directly from the probabilities, without calculating expected frequencies at all.

x_i	$p(x_i)$	$x_i p(x_i)$
1	$\dfrac{1}{6}$	$1 \times \dfrac{1}{6}$
2	$\dfrac{1}{6} \times \dfrac{5}{6}$	$2 \times \dfrac{1}{6} \times \dfrac{5}{6}$
3	$\dfrac{1}{6} \times \left(\dfrac{5}{6}\right)^2$	$3 \times \dfrac{1}{6} \times \left(\dfrac{5}{6}\right)^2$
4	$\dfrac{1}{6} \times \left(\dfrac{5}{6}\right)^3$	$4 \times \dfrac{1}{6} \times \left(\dfrac{5}{6}\right)^3$
5	$\dfrac{1}{6} \times \left(\dfrac{5}{6}\right)^4$	$5 \times \dfrac{1}{6} \times \left(\dfrac{5}{6}\right)^4$
$\vdots$	$\vdots$	$\vdots$
n	$\dfrac{1}{6} \times \left(\dfrac{5}{6}\right)^{n-1}$	$n \times \dfrac{1}{6} \times \left(\dfrac{5}{6}\right)^{n-1}$

Thus,

$$\sum_i x_i p(x_i) = 1 \times \frac{1}{6} + 2 \times \frac{1}{6} \times \frac{5}{6} + 3 \times \frac{1}{6} \times \left(\frac{5}{6}\right)^2 + 4 \times \frac{1}{6} \times \left(\frac{5}{6}\right)^3 + 5 \times \frac{1}{6} \times \left(\frac{5}{6}\right)^4 + \dots$$

The sum of these terms forms a series but not a simple geometric series with a common ratio.

The main difficulty is caused by the whole number coefficients and we shall use a technique to remove these. Here is our sum of terms again with 5/6 times that series written underneath:

$$\bar{x} = 1 \times \frac{1}{6} + 2 \times \frac{1}{6} \times \frac{5}{6} + 3 \times \frac{1}{6} \times \left(\frac{5}{6}\right)^2 + 4 \times \frac{1}{6} \times \left(\frac{5}{6}\right)^3 + 5 \times \frac{1}{6} \times \left(\frac{5}{6}\right)^4 + \dots$$

$$\frac{5}{6}\bar{x} = \qquad 1 \times \frac{1}{6} \times \frac{5}{6} + 2 \times \frac{1}{6} \times \left(\frac{5}{6}\right)^2 + 3 \times \frac{1}{6} \times \left(\frac{5}{6}\right)^3 + 4 \times \frac{1}{6} \times \left(\frac{5}{6}\right)^4 + \dots$$

We can now subtract this second series from the first series:

$$\frac{1}{6}\bar{x} = \frac{1}{6} \quad + \frac{1}{6} \times \frac{5}{6} \quad + \frac{1}{6} \times \left(\frac{5}{6}\right)^2 \quad + \frac{1}{6} \times \left(\frac{5}{6}\right)^3 \quad + \frac{1}{6} \times \left(\frac{5}{6}\right)^4 \quad + \dots$$

This is a geometric series with

$$a = \frac{1}{6}$$

$$r = \frac{5}{6}$$

Its sum is given by $a/(1-r)$ so that

$$\frac{1}{6}\bar{x} = \frac{1}{6} \bigg/ \left(1 - \frac{5}{6}\right)$$

$$\frac{1}{6}\bar{x} = 1$$

$$\bar{x} = 6$$

Alternatively, you may notice that

$$\frac{1}{6}\bar{x} = \sum_i p(x_i)$$

$$\frac{1}{6}\bar{x} = 1$$

$$\bar{x} = 6$$

Thus the theoretical value for the mean is 6.

Our experimental value for $\bar{x}$ was close to this at 6.475 even with only 122 trials.

Practical 2.1 Revisited

In a Series of Throws of a Coin, Which is More Likely to Occur First, HHH or THT?

If only three coins are tossed, these two outcomes are equally likely, each occurring with a probability of $1/8$. However, is that true if we look to see which occurs first in a sequence of throws?

Practical Follow-up 2.9

Find the class results for Practical 2.1. Go through the records of all the trials and underline which ever comes first, HHH or THT. When you find one or other of these two events, regard the next throw as the beginning of a new sequence. Our results are given at the end of this chapter.

We found 30 occurrences of HHH first and 33 occurrences of THT first. Note that HHHHTT is one occurrence of HHH (not two!) because the next sequence begins at HTT.

How do your results compare with ours? Did you find more occurrences of THT first? Or do you think that the two outcomes are equally likely?

Theoretical Probabilities

Let

THT = event A
HHH = event B

First Throw

At the start, the player obtains either H or T (each with a probability of $1/2$) (Figure 2.21).

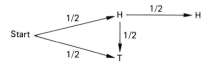

Figure 2.21 Probabilities on the first throw

Second Throw

Looking at the second throw, we shall consider the player who has H on the first throw. If he gets heads again, he can progress towards HHH (event B). If he gets tails, he now cannot get HHH but this tail could be considered as the first working towards a sequence of THT (Figure 2.22).

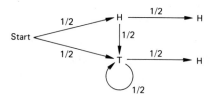

Figure 2.22 Probabilities on the second throw

Next we look at the possibilities for a player obtaining tails first time. If he gets TH, he can progress towards THT (event A). If he gets TT, his first tail is useless now but the second tail can be regarded as the possible beginning of a THT sequence (Figure 2.23).

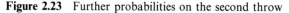

Figure 2.23 Further probabilities on the second throw

Third Throw

From HH the player may get a third H and then complete event B but, if he gets HHT, his first two throws are useless and he goes right back to the first tail for a possible sequence of THT.

From TH the player could get THH in which case he has two heads to work towards event B (HHH). He might get THT, however, and arrive at the completion of event A (Figure 2.24).

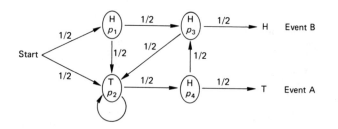

Figure 2.24 Probabilities on the third throw

Let p_1, p_2, p_3 and p_4 be the probabilities of eventually reaching event A from the points shown on the diagram. $P(A)$ denotes the probability of reaching even A from the start.

At the start, event A can be reached via point 1 or point 2:

$$P(A) = \frac{1}{2}p_1 + \frac{1}{2}p_2$$

From point 1, we can progress to either point 2 or point 3:

$$p_1 = \frac{1}{2}p_3 + \frac{1}{2}p_2$$

From point 2, we can go on to point 4 or stay at point 2:

$$p_2 = \frac{1}{2}p_4 + \frac{1}{2}p_2$$

From point 3, we can move on to event B (making it impossible then to reach event A first) or we can go to point 2:

$$p_3 = \frac{1}{2} \times 0 + \frac{1}{2}p_2$$

From point 4, we can reach event A (probability, $1/2$) or go on to p_3:

$$p_4 = \frac{1}{2} + \frac{1}{2}p_3$$

To find $P(A)$, go back to the expression for p_2:

$$p_2 = \frac{1}{2}p_4 + \frac{1}{2}p_2$$

Substitute in for p_4:

$$p_2 = \frac{1}{4} + \frac{1}{4}p_3 + \frac{1}{2}p_2$$

Substitute in for p_3:

$$p_2 = \frac{1}{4} + \frac{1}{8}p_2 + \frac{1}{2}p_2$$

$$p_2 = \frac{1}{4} + \frac{5}{8}p_2$$

$$\frac{3}{8}p_2 = \frac{1}{4}$$

$$p_2 = \frac{1}{4} \times \frac{8}{3}$$

$$p_2 = \frac{2}{3}$$

From this value we find that

$$p_3 = \frac{1}{3}$$

$$p_1 = \frac{1}{6} + \frac{1}{3} = \frac{1}{2}$$

and now we can find $P(A)$:

$$P(A) = \frac{1}{2}p_1 + \frac{1}{2}p_2$$

$$= \frac{1}{4} + \frac{1}{3}$$

$$= \frac{7}{12}$$

Thus, the probability of reaching event A (THT) first is 7/12, while the probability of reaching event B (HHH) first is 5/12.

This supports our empirical results and our totals are very close to those predicted by the theoretical probabilities (Table 2.8).

Table 2.8 Experimental and theoretical frequencies

Event	Experimental frequency	Theoretical frequency
THT	33	36
HHH	30	25

Section 2.7 How to Discover the Truth in Tricky Situations

Practical 2.10

You require 2 dice of different colours, e.g. one red die and one white die.

In this section we describe how to use a technique known as *randomised response*. This can be used to find out the proportion of people in a group who take part in an unacceptable (or even illegal) activity, without any individual having to reveal their involvement directly. An American survey in the 1960s used the randomised response technique to find out how many young women had had illegal abortions (the question must have a yes/no answer).

Before you begin to think of suitable questions of an embarrassing nature, it may be better to try out the technique using a simple question such as 'Are you a boy?' or 'Are you 17?' which you can check afterwards. You will be able to see then whether the technique gives you a close estimate of the actual number of boys in the class.

Each student must roll 2 dice of different colours, one red and one white say. On the trial run, everyone may see the results of each person's throw of the dice and you can help them to decide on their answer.

1. If you score an *odd* number on the *red* die, you must answer the survey question truthfully (and ignore the white die).
2. If you score an *even* number on the *red* die, look at the white die. Answer the question

'Did you obtain an odd number on the white die?' So, if the *white* die is *odd*, answer yes and, if the *white* die is *even*, answer no.

When this scheme is used properly, respondents may answer yes because they are answering the survey question or yes because they are answering about the white die. No one will be able to identify the true cause for their answer.

Table 2.9 shows examples so that you can see how the scheme works. The survey question used is 'Are you a boy?'

Table 2.9   Example responses

	Red score	White score		Yes/no
Jilly	5	1	Answers survey truthfully	No
John	3	2	Answers survey truthfully	Yes
Peter	2	3	Answers about white die	Yes
Roger	4	4	Answers about white die	No

Now use the scheme yourself using a simple trial question which can be easily checked, e.g. 'Are you a boy?', 'Are you 17?' or 'Have you started driving lessons?'.

Your Results

Since rolling a die is a random event, we expect the score on the red die to be odd half the time and even half the time. This means that about half the class should answer the survey question and the other half answer about the white die (Figure 2.25).

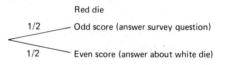

Figure 2.25   Results on rolling the red die

For those answering about the white die, we expect half to answer yes and half to answer no, according to their score on the die (Figure 2.26).

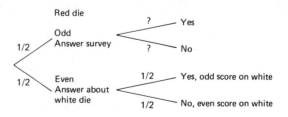

Figure 2.26   Results on answering survey question

When a class of 28 students used this trial, 18 gave yes answers and 10 no answers. The survey question was 'Are you a boy?'. How many boys do we think there should be? Out of 28 students, about 14 should be answering the survey question and 14 should be answering about the white die.

First we need to eliminate the white die answers, so that we can concentrate on the answers to the survey question.

Of our 14 answers about the white die, 7 should be yes and 7 should be no (Figure 2.27).

Subtracting 7 from 18, this leaves 11 survey *yes* answers (and similarly we have 3 survey *no* answers). Only the survey answers can be used to estimate the number of boys in the class. Out of 14 survey answers then, 11 say yes. So we estimate the proportion of boys to be 11/14, giving 22 out of 28 in the whole class.

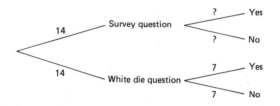

Figure 2.27 Results on rolling the white die

Work through this analysis with your answers. How many boys does this technique give as an estimate? Your answer is only an estimate; it is not likely that exactly half the scores on either die will be odd. However, with a large number of people answering a question using this technique, the estimate can be very close, and there is no reason for anyone not to tell the truth even to sensitive survey questions.

Practical 2.11

Now use the randomised response technique on your class using a more controversial question. Alternatively, ask whether you can try out this technique on a class of younger children and ask the question, 'Do you smoke?'. Do not forget that each person must keep their scores on the dice secret. Also, younger children may find the instructions difficult to follow. It will help to have the instructions written on the blackboard and you may have to explain them several times!

Buffon's Needle

Proof is given that the probability that a pin lands on a line is $2/\pi$ (provided that the lines are the same distance apart as the length of the pin).

Let the distance between lines be d and the length of a pin d (Figure 2.28). We shall call the distance from the centre of the pin to the nearest line y.

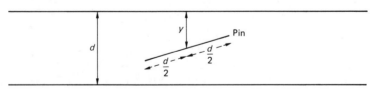

Figure 2.28 Lengths in Buffon's needle experiment (Practical 2.2)

It is easy to see that y can be any value between 0 and $d/2$ and, if the pin falls at random, then each value of y is equally likely.

We must also consider the angle of the pin to the horizontal (Figure 2.29). This angle θ may be of any magnitude between 0 and π. Again, if the pin falls at random, then each

Figure 2.29 Angle in Buffon's needle experiment (Practical 2.2)

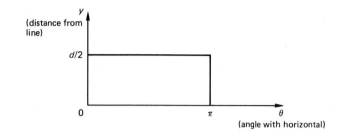

Figure 2.30 Total sample space

angle is equally likely. Thus the total sample space may be represented by the area shown in Figure 2.30.

Any possible result (for the distance of the pin and its angle) may be represented by a point inside this sample space.

To find the probability that a pin crosses a line, we need to consider those that do in mathematical terms (Figure 2.31).

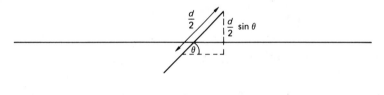

Figure 2.31 Pin crossing a line

The vertical distance from the centre of the pin to its tip is $(d/2)\sin\theta$. The pin will cross a line if

$$y < \frac{d}{2}\sin\theta$$

We can draw the function $y = (d/2)\sin\theta$ inside the sample space (Figure 2.32). For

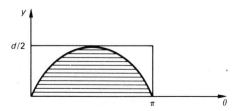

Figure 2.32 The function $y = (d/2)\sin\theta$ inside the sample space

$P(Y < (d/2)\sin\theta)$, we require the shaded area as a proportion of the total area:

$$\text{shaded area} = \int_0^\pi \frac{d}{2}\sin\theta\,d\theta$$

$$= \frac{d}{2}\left[-\cos\theta\right]_0^\pi$$

$$= \frac{d}{2} \times -(-1-1)$$

$$= \frac{d}{2} \times 2$$

$$= d$$

Now

$$\text{total area} = \pi\frac{d}{2}$$

and, therefore,

$$\frac{\text{shaded area}}{\text{total area}} = \frac{d}{\pi(d/2)}$$

Thus,

$$P\left(Y < \frac{d}{2}\sin\theta\right) = \frac{2}{\pi}$$

Results of Tossing a Coin 400 Times

Table 2.10   Tossing a coin

```
H T H T T H H H T H T H T H T T H T T T T T T H T H H H H T T T T T T H H T
H H H T H H H H T H H H T T T T T H T T T T T H H H H H T H T T T H H H T T
H H H T H T T H H H T H H T T H H H H H T T T T T T H T T H H H H H T H T
H H H H H T H H T H H H T T T H H T H T H H H T H H H T T H T H T T H H T T H
H H T H H T T H T T T H H H T T T T H T T H T T H H H T H H H T H T T H T H H
H T H H T T T T H H H T H H T H H T H H T T H H H T T H T H T H H H H T H H
H H H T T H T T H T T T H H T H H H T T H H T T H H H H T H T H T T H T H H
H T H T T T H T T H H T H H H T H T T H T T H T T T T T T H H T T T T T T
H H T H H H H H H T H T H T T T H T T H H H T H H T T H H T H T H H H T T H
H T T T T T H H T H H H H T T T T T H T H T H H T H H T H T H H T H T T T T
H
```

Chapter 3 The Binomial Distribution

Section 3.1 The 'Frivolous Pursuit' Quiz

Practical 3.1 The Quiz

Try out our seven-question quiz among your own class and then also with other groups of students in order to collect together a minimum of 50 sets of results.

The quiz questions are reproduced at the end of this chapter.

Arrange to have the papers marked (the correct answers appear in Section 3.7) and record the number of correct answers for each person. Make a frequency table for the number of correct answers (from 0 to 7) for the entire group of students answering the quiz.

Our Results

We tried out the final quiz on 104 students and Table 3.1 is the frequency table for the number of correct answers (out of 7) gained by those students. Is this frequency distribution as we would expect if our students were guessing the answers?

Table 3.1 Answers to the quiz

Number of correct answers out of 7	Frequency (number of students)
0	0
1	8
2	26
3	28
4	30
5	8
6	4
7	0
Total	104

Section 3.2 Finding a Probability Model

Since it is unlikely that the answers will be known, we expect each student to guess correctly with a probability of 1/2, for each question (Figure 3.1).

Figure 3.1 Probabilities of guessing the correct answer to 1 question

For 2 questions, we would expect 1/4 of our students to guess both answers correctly, 1/2 to guess one answer correctly and 1/4 to get both wrong (Figure 3.2). Note that there are two outcomes giving one correct answer, as we can select either one of two positions for that correct answer—either on the first question or on the second.

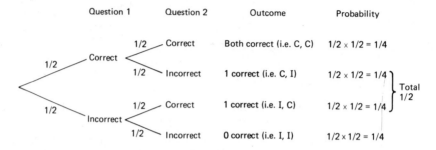

Figure 3.2   Probabilities of guessing the correct answers to 2 questions

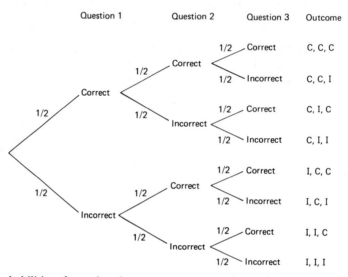

Figure 3.3   Probabilities of guessing the correct answers to 3 questions

We can extend our model to 3 questions (Figure 3.3). The possible outcomes are

0 correct answers   1 outcome

I, I, I   probability $\frac{1}{8}$

1 correct answer   3 outcomes

C, I, I

I, C, I $\Big\}$ probability $\frac{3}{8}$

I, I, C

2 correct answers   3 outcomes

C, C, I

C, I, C $\Big\}$ probability $\frac{3}{8}$

I, C, C

3 correct answers   1 outcome

C, C, C   probability $\frac{1}{8}$

Note that there are 3 outcomes for 1 correct answer as we can select any one of 3 positions for it. The same is true for 1 incorrect answer.

If we consider more than 3 questions, it becomes more difficult to draw a tree diagram, but the possible outcomes for 4 questions are given in Table 3.2.

Table 3.2   The possible outcomes for 4 questions

0 correct answers	1 correct answer	2 correct answers	3 correct answers	4 correct answers
1 outcome	4 outcomes	6 outcomes	4 outcomes	1 outcome
I I I I	C I I I I C I I I I C I I I I C	I I C C C C I I I C I C C I C I I C C I C I I C	I C C C C I C C C C I C C C C I	C C C C

To count the number of outcomes for each event, we can use combinations as we are selecting positions for correct and incorrect answers (Table 3.3). For example, if we wish to calculate the number of ways of choosing 3 of the 4 questions to label 'correct', we need the number of selections of 3 items from 4. This is written either as $^4C_3$ or $\binom{4}{3}$ and is calculated as $4!/1!3!$

Table 3.3   Calculating the number of outcomes for 4 questions

	Number of outcomes	
0 correct answers	$^4C_0$	$= 1$
1 correct answer	$^4C_1$	$= 4$
2 correct answers	$^4C_2 = \frac{4 \times 3}{2 \times 1}$	$= 6$
3 correct answers	$^4C_3 = \frac{4 \times 3 \times 2}{3 \times 2 \times 1}$	$= 4$
4 correct answers	$^4C_4 = \frac{4 \times 3 \times 2 \times 1}{4 \times 3 \times 2 \times 1} = 1$	
Total	16	

As correct and incorrect answers are equally likely, the associated probabilities are $P(0) = 1/16$, $P(1) = 4/16$, $P(2) = 6/16$, $P(3) = 4/16$ and $P(4) = 1/16$. (In general, to calculate the number of outcomes for r successes in n trials, we need ^nC_r or $\binom{n}{r}$ calculated as $n!/(n-r)!r!$ and defining $0! = 1$.)

A Probability Model for 7 Questions

By using this idea of selecting positions for correct (or incorrect) answers, we can predict the pattern for the numbers of correct answers that we could expect for 7 questions (Table 3.4).

Table 3.4   Outcomes for 7 questions

	Number of outcomes		Probability
0 correct answers	$^{7}C_{0}$	$= 1$	$\dfrac{1}{128}$
1 correct answer	$^{7}C_{1}$	$= 7$	$\dfrac{7}{128}$
2 correct answers	$^{7}C_{2} = \dfrac{7 \times 6}{2 \times 1}$	$= 21$	$\dfrac{21}{128}$
3 correct answers	$^{7}C_{3} = \dfrac{7 \times 6 \times 5}{3 \times 2 \times 1}$	$= 35$	$\dfrac{35}{128}$
4 correct answers	$^{7}C_{4} = \dfrac{7 \times 6 \times 5 \times 4}{4 \times 3 \times 2 \times 1}$	$= 35$	$\dfrac{35}{128}$
5 correct answers	$^{7}C_{5} = \dfrac{7 \times 6 \times 5 \times 4 \times 3}{5 \times 4 \times 3 \times 2 \times 1}$	$= 21$	$\dfrac{21}{128}$
6 correct answers	$^{7}C_{6} = \dfrac{7 \times 6 \times 5 \times 4 \times 3 \times 2}{6 \times 5 \times 4 \times 3 \times 2 \times 1}$	$= 7$	$\dfrac{7}{128}$
7 correct answers	$^{7}C_{7} = \dfrac{7 \times 6 \times 5 \times 4 \times 3 \times 2 \times 1}{7 \times 6 \times 5 \times 4 \times 3 \times 2 \times 1} = 1$		$\dfrac{1}{128}$
Total		128	$\dfrac{128}{128}$

We can compare our actual results with those predicted by the probability model (Table 3.5). To obtain expected frequencies for each number of correct answers, multiply each theoretical probability by the total number of people in the group. We have 104 students in our group; so we multiply each probability by 104.

As you can see from the line graph in Figure 3.4, our model predicts results fairly close to those we actually obtained. Our experiment, however, did produce more people with 2 correct answers than predicted, although that group does include some students who wrote false for every answer. It could perhaps be argued that these people did not answer the test properly and have distorted our results. If we were to eliminate those (and also anyone who answered true for every answer), we would get a better fit between the theoretical model and the actual frequencies. However, it is debatable whether we are really justified in taking that course of action.

Table 3.5 Expected frequencies for each number of correct answers

Number x of correct answers	Theoretical probability $p(x)$	Expected frequencies $104p(x)$	Experimental frequencies (our results)
0	$\dfrac{1}{128}$	0.8	0
1	$\dfrac{7}{128}$	5.7	8
2	$\dfrac{21}{128}$	17.1	26
3	$\dfrac{35}{128}$	28.4	28
4	$\dfrac{35}{128}$	28.4	30
5	$\dfrac{21}{128}$	17.1	8
6	$\dfrac{7}{128}$	5.7	4
7	$\dfrac{1}{128}$	0.8	0

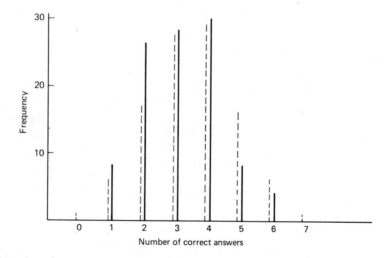

Figure 3.4 Number of correct answers (out of 7): ———, experimental frequencies; – – –, theoretical frequencies as predicted by a binomial model

Practical follow-up **3.1**

1. Calculate the expected frequencies for your group size, and show your actual frequencies and expected frequencies together on a line graph.
2. **The Examiner's Problem**. If an examiner sets 10 questions each with 2 alternative answers

(one right and one wrong), what is the probability that a candidate gains a pass mark of 50% purely by guesswork?

Section 3.3   The General Binomial Probability Model

In the example that we have used so far, the probability of a correct (and also an incorrect) answer is 1/2, and our resulting probability distribution is always symmetrical.

However, this need not be so. Look at the following question.

Does the world's most famous bell, Big Ben in the clock tower of the House of Commons in London, have a mass of 12.575 tons, 13.05 tons, 13.575 tons or 14.25 tons?

Here there are 4 alternative answers, and only one is correct; so the probability of guessing a correct answer (usually called p) is 1/4, and the probability of guessing an incorrect answer ($1 - p$; usually called q) is 3/4.

For a series of 5 questions of this type, we can calculate the probabilities of 0, 1, 2, 3, 4 or 5 correct answers provided that we now take into account the fact that $p = 1/4$ and $q = 3/4$. We still use combinations to calculate the number of possible outcomes (Table 3.6).

Table 3.6   Probabilities of correct answers for 5 questions

Number x of correct answers	Number of outcomes $\times$		probability of each outcome	Probability
0	$^5C_0 = 1$	$\times$	$\left(\dfrac{3}{4}\right)^5$	$= 1 \times \dfrac{243}{1024} = \dfrac{243}{1024}$
1	$^5C_1 = 5$	$\times$	$\left(\dfrac{3}{4}\right)^4\left(\dfrac{1}{4}\right)$	$= 5 \times \dfrac{81}{1024} = \dfrac{405}{1024}$
2	$^5C_2 = \dfrac{5 \times 4}{2 \times 1}$	$\times$	$\left(\dfrac{3}{4}\right)^3\left(\dfrac{1}{4}\right)^2$	$= 10 \times \dfrac{27}{1024} = \dfrac{270}{1024}$
3	$^5C_3 = \dfrac{5 \times 4 \times 3}{3 \times 2 \times 1}$	$\times$	$\left(\dfrac{3}{4}\right)^2\left(\dfrac{1}{4}\right)^3$	$= 10 \times \dfrac{9}{1024} = \dfrac{90}{1024}$
4	$^5C_4 = \dfrac{5 \times 4 \times 3 \times 2}{4 \times 3 \times 2 \times 1}$	$\times$	$\left(\dfrac{3}{4}\right)\left(\dfrac{1}{4}\right)^4$	$= 5 \times \dfrac{3}{1024} = \dfrac{15}{1024}$
5	$^5C_5 = \dfrac{5 \times 4 \times 3 \times 2 \times 1}{5 \times 4 \times 3 \times 2 \times 1}$	$\times$	$\left(\dfrac{1}{4}\right)^5$	$= 1 \times \dfrac{1}{1024} = \dfrac{1}{1024}$

The Examiner's Problem Solved (?)

The examiner decides that too many of the candidates can pass his previous test by guessing the answers. He sets a new test again with 10 questions, but now with 4 alternative answers to each question. What is the probability that a candidate can pass the new test by guesswork? (The pass mark is still 50%.)

The General Binomial Model

Our probability model can be adapted to fit any situation in which there are, firstly, a fixed number of trials and, secondly, two alternative outcomes to each trial (e.g. correct answer or incorrect answer), which can be labelled as success or failure, and constant probability of success. In situations in which there are more than 2 outcomes, the binomial model can be used provided the outcomes can be reclassified as success or failure. For example, if a die is thrown, a score of 6 can be classified as success and scores of 1, 2, 3, 4 or 5 as failure.

The constant probability of a success at each trial is known as p and the probability of failure is known as q. For a series of n repeats of the trial (e.g. 10 test questions), the probabilities of 0, 1, 2, 3, ... up to n successes are given by successive terms of the expansion of $(q + p)^n$.

The probability distribution in Table 3.7 is known as the binomial distribution and can be written in an abbreviated form as $B(n, p)$. Thus $B(10, 1/4)$ means a binomial distribution with $n = 10$ and $p = 1/4$.

Table 3.7 Binomial distribution

Number r of successes	Probability of r successes	
	= number of outcomes	$\times$ probability of 1 outcome
0	$^nC_0 = 1$	$\times \ q^n$
1	$^nC_1 = n$	$\times \ q^{n-1}p$
2	$^nC_2 = \dfrac{n(n-1)}{2!}$	$\times \ q^{n-2}p^2$
3	$^nC_3 = \dfrac{n(n-1)(n-2)}{3!}$	$\times \ q^{n-3}p^3$
4	$^nC_4 = \dfrac{n(n-1)(n-2)(n-3)}{4!}$	$\times \ q^{n-4}p^4$
$\vdots$	$\vdots$	$\vdots$
n	$^nC_n = 1$	$\times \ p^n$

The general term of a binomial probability distribution is given as

$$P(X = r) = {}^nC_r p^r q^{n-r} \quad \text{for} \quad r = 0, 1, 2, ..., n$$

Section 3.4 Alternative Practicals

Each of the following practicals should give results which fit a binomial model. In order to calculate theoretical probabilities, you may need to estimate p from your experimental results.

p equals probability of a success and, if not known, may be estimated from

$$\frac{\text{total number of successes in your experiment}}{\text{total number of trials in your experiment}}$$

Practical 3.2a Distribution of Girls in Two-child Families

Find the distribution of girls in 50–100 two-child families and make a frequency table such as Table 3.8.

If a Binomial model is to be useful in this situation what is the value of n?

Compare these frequencies with the expected frequencies calculated for $p = 1/2$ and again from p estimated from your sample. Which of the two values of p do you think is the better value to use?

Table 3.8   Number of girls in two-child families

Number of girls	Frequency
0	
1	
2	
Total	

Practical 3.2b   Distribution of Girls in Three-child Families

Find the distribution of girls in 50–100 three-child families and compare your actual frequencies with the expected theoretical frequencies.

Practical 3.3   Proportion of Cress Seeds Which Germinate

Plant cress seeds in groups of 4 on damp blotting paper on plastic or polystyrene trays. It would be useful to have results for 50–100 groups of 4; so lines can be drawn in biro on the blotting paper in each tray to divide it up into several groups. Count the number of seeds in each group of 4 which germinate (Table 3.9). For this experiment $n = 4$ and p, the proportion of seeds which germinate, will need to be estimated from your results:

$$p = \frac{\text{number of seeds which germinate (total)}}{\text{number of seeds planted (total)}}$$

$$p = \frac{\sum fx}{\left(\sum f\right) \times 4}$$

Calculate the theoretical expected frequencies and compare them with your results.

Table 3.9   Number of seeds which germinate

Number x of seeds which germinate	Frequency	fx
0		
1		
2		
3		
4		
Total		

Practical 3.4   Average Contents in Matchboxes

Supermarkets often sell matchboxes in packets of 6. To do this experiment, you need to buy 50–100 packets of 6. Since this will take some time, get someone to help you.

For each packet of 6 boxes, count the number of boxes which contain less than the stated average contents (printed on the box). Record your results on a frequency table such as Table 3.10.

Table 3.10   Average contents of matchboxes

Number x of boxes containing less than the average contents	Frequency f	fx
0		
1		
2		
3		
4		
5		
6		
Total	$\sum f =$	$\sum fx =$

In order to calculate theoretical expected frequencies, you will need to estimate p, the probability that a box contains less than the stated average:

$$p = \frac{\text{number of boxes containing less than the stated average}}{\text{total number of matchboxes}}$$

$$p = \frac{\sum fx}{\left(\sum f\right) \times 6}$$

Practical 3.5   Left-handed Pupils

At lunch time, count the number of left-handed pupils at each dinner table. (The tables must all seat the same number of people so that $n = 6$ (say) or $n = 8$; ignore tables that do not have this number of people at them.) See whether your results are close to those for a binomial distribution.

Practical 3.6   Average Weights of Packets of Cereal

If you have contact with a friendly shopkeeper, you may be able to conduct this experiment outside shop-opening hours. Most foodstuffs in tins or packets have an average weight printed on them by the symbol e. This means that the weights should form a distribution with that average, but half may be under that weight (and half over). For each carton of 10 packets of cereal (say), find out how many are over the stated average weight. You will not be able to empty each package; so you will have to weigh just one empty packet so that you can subtract that amount from each weight. Compare your results with those predicted by a binomial distribution for the following.

1. $p = 1/2$.
2. p estimated from your sample.

Section 3.5   The Mean and Variance of a Binomial Distribution

From our experimental results for Practical 3.1 (the Frivolous Pursuit quiz), we have calculated the mean and variance of the number of correct answers scored (Table 3.11).

Table 3.11   Experimental results

Number x of correct answers	Frequency x	fx	fx^2
0	0	0	0
1	8	8	8
2	26	52	104
3	28	84	252
4	30	120	480
5	8	40	200
6	4	24	144
7	0	0	0
Total	104	328	1188

Therefore,

$$\text{mean } \bar{x} = \frac{\sum fx}{\sum f}$$

$$\bar{x} = \frac{328}{104} = 3.154$$

$$\text{variance} = \frac{\sum fx^2}{\sum f} - \left(\frac{\sum fx}{\sum f}\right)^2$$

$$= \frac{1188}{104} - \left(\frac{328}{104}\right)^2$$

$$= 11.423 - 9.947$$

$$= 1.476$$

Practical Follow-up 3.2

Calculate the mean and variance from your frequency table of experimental results. If your results are for the Frivolous Pursuit quiz, how do your mean and variance compare with ours?

Theoretical Values for the Mean and Variance

In Chapter 2, we showed that the mean of a discrete probability distribution can be found from

$$\mu = \sum_{x=1}^{n} xp(x)$$

and the variance from

$$\sigma^2 = \sum_{x=1}^{n} (x - \mu)^2 p(x)$$

or

$$\sigma^2 = \sum_{x=1}^{n} x^2 p(x) - \left(\sum_{x=1}^{n} xp(x)\right)^2$$

We can compare our experimental values for the mean and variance with the theoretical values calculated from the binomial probabilities.

Theoretical Binomial Probability Distribution

The theoretical values for the binomial probability distribution are given in Table 3.12.

Table 3.12   Theoretical binomial distribution for $n = 7$, $p = \frac{1}{2}$

Number x of correct answers	$p(x)$	$xp(x)$	$x^2p(x)$
0	$\dfrac{1}{128}$	0	0
1	$\dfrac{7}{128}$	$\dfrac{7}{128}$	$\dfrac{7}{128}$
2	$\dfrac{21}{128}$	$\dfrac{42}{128}$	$\dfrac{84}{128}$
3	$\dfrac{35}{128}$	$\dfrac{105}{128}$	$\dfrac{315}{128}$
4	$\dfrac{35}{128}$	$\dfrac{140}{128}$	$\dfrac{560}{128}$
5	$\dfrac{21}{128}$	$\dfrac{105}{128}$	$\dfrac{525}{128}$
6	$\dfrac{7}{128}$	$\dfrac{42}{128}$	$\dfrac{252}{128}$
7	$\dfrac{1}{128}$	$\dfrac{7}{128}$	$\dfrac{49}{128}$

Therefore,

$$\sum xp(x) = \frac{448}{128}$$

$$\sum x^2 p(x) = \frac{1792}{128}$$

$$\mu = \sum xp(x)$$

$$\mu = \frac{448}{128} = 3.5$$

$$\sigma^2 = \sum x^2 p(x) - \left(\sum xp(x)\right)^2$$

$$\sigma^2 = \frac{1792}{128} - \left(\frac{448}{128}\right)^2$$

$$\sigma^2 = 14 - 12.25$$

$$\sigma^2 = 1.75$$

The mean of a binomial distribution may also be calculated (very quickly) from the formula

$$\mu = np$$

and the variance from

$$\sigma^2 = npq$$

A proof is given later in the section.

For this example with $n = 7$, $p = 1/2$ and $q = 1/2$,

$$\mu = 7 \times \frac{1}{2}$$

$$\mu = 3.5$$

and

$$\text{Var}(x) = 7 \times \frac{1}{2} \times \frac{1}{2}$$

$$= 1.75$$

Practical Follow-up 3.3   (**Alternative Practicals**)

If you have results for any of the alternative practicals (Section 3.4), calculate the theoretical values for the mean and variance and compare them with the values calculated from your experimental results.

Proof of the Mean and Variance of the Binomial Distribution

For independent random variables,

$$E(X + Y) = E(X) + E(Y)$$

and

$$\text{Var}(X + Y) = \text{Var}(X) + \text{Var}(Y)$$

The binomial distribution can be considered as the sum of n independent Bernoulli trials. At each trial

$$p(\text{success}) \text{ of } P(X = 1) = p$$

$$p(\text{failure}) \text{ or } P(X = 0) = q$$

For one Bernoulli trial,

$$E(X) = \sum xp(x)$$

$$= (1 \times p) + (0 \times q) = p$$

$$\text{Var}(X) = \sum x^2 p(x) - (\text{mean})^2$$

$$= 1 \times p + 0 \times q - p^2$$

$$= p - p^2$$

$$= p(1 - p)$$

$$= pq$$

For the sum of n independent Bernoulli trials,

$$E\left(\sum_{i=1}^{n} X\right) = nE(X) = np$$

$$\text{Var}\left(\sum_{i=1}^{n} X\right) = \sum_{i=1}^{n} \text{Var}(X) = npq$$

Section 3.6 An Industrial Application of the Binomial Distribution

Whether it be producing engines, nails or currant buns, any manufacturing process will produce some substandard or defective items. All manufacturers test their output; most cannot afford to test every item and so will take out a sample for testing. If items are produced in batches, a few items from each batch will be tested and the whole batch will be accepted or rejected on the basis of this small sample.

In addition to this checking at the end of the process, the manufacturer will wish to ensure that his raw materials are of the standard that he requires. Technicians supervising deliveries of raw materials take samples from each load as it arrives. They cannot examine or test the entire load, and so have to accept or reject the load on the basis of a few results.

Here is an example of this type of sampling, operated by a confectionery manufacturers. One of their main raw materials is, of course, sugar, which is delivered by lorry in large sacks. The technicians supervising the delivery will not wish to accept poor-quality sugar (and in particular damp sugar) as this could produce an entire batch of 'reject' sweets. However, the firm have built up long-standing relationships with their suppliers and will not wish to send back a lorry load of sugar which does in fact meet the agreed standards.

Here is the sampling scheme used to test loads of sugar coming in to the factory. The scheme has been devised and modified over time so that the technical staff are reasonably happy that they are able to detect substandard loads without being overcautious. They are also helped by the fact that corrections can be made (within limits) during the manufacturing process to adjust the viscosity of the sweet mix.

At stage 1, two sacks are sampled from the lorry load. If both samples are up to standard, that load is accepted straight away without any further testing. If both samples are too damp, then the lorry is sent back immediately.

If the stage-1 sampling produces one bad sample, then the technician takes two further samples. This is known as stage 2 of this two-stage sampling scheme. At stage 2, the batch will only be accepted if *both* samples are satisfactory.

The entire scheme can be represented by a tree diagram as in Figure 3.5. This diagram shows the probability of accepting a load of sugar given that 10% of the sacks contain substandard sugar.

Of all the outcomes possible, only either of the following leads to the acceptance of a load of sugar.

1. The load is accepted at stage 1 (with 10% of sacks substandard), i.e. $p = 0.9 \times 0.9 = 0.81$.
2. Stage-2 samples are taken (i.e. one stage-1 sample was good and one was bad), i.e. $p = 0.9 \times 0.1 \times 2 = 0.18$, *and* both stage-2 samples are good, i.e. $p = 0.9 \times 0.9 = 0.81$, giving $0.18 \times 0.81 = 0.1458$.

The total probability of accepting a load in which 10% of the sacks are substandard is $0.81 + 0.1458 = 0.9558$.

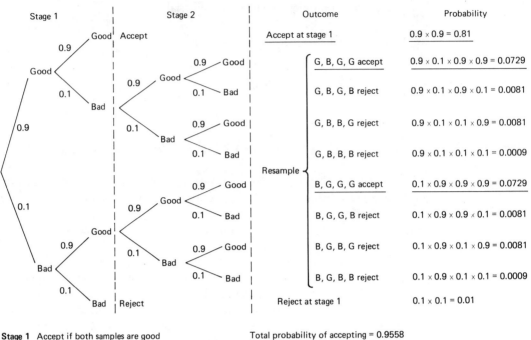

Stage 1 Accept if both samples are good
Reject if both samples are bad
Resample if 1 is good and 1 is bad

Stage 2 Only accept if both are good

Total probability of accepting = 0.9558
(given that 10% of load is substandard)

Figure 3.5 Two-stage sampling of sugar

This calculation can be written in general terms for any given proportion of sacks containing bad sugar in a load.

Let p be the proportion of sacks containing good sugar and q the proportion of sacks containing bad sugar; then either of the following leads to the acceptance of a load of sugar.

1. p (accepting at stage 1) $= p^2$.
2. p (of taking stage-2 samples) $= 2pq$,
   and p (both stage-2 samples are good) $= p^2$,
   giving p (of accepting load at stage 2) $= 2pq \times p^2 = 2p^3q$.

Therefore,

total probability of accepting a load $= p^2 + 2p^3q$

The probability of accepting a load can be calculated for various values of p and q as shown in Table 3.13.

Probabilities of Accepting Loads with Different Proportions of Sacks Containing Good Sugar

Let the proportion of sacks containing good sugar be p and the proportion of sacks containing bad sugar q. The probability of accepting a load of sugar can be plotted against the proportion q of sacks containing bad sugar in a load, to give the operating characteristic curve (Figure 3.6).

This curve shows that, even for a 70% proportion of bad sugar, the probability of accepting a load is quite high (nearly 0.13). While the manufacturers are obviously satisfied with the sampling scheme (otherwise they would change it), it does not seem very efficient.

Table 3.13 Probability of accepting a load

p	q	Accept at stage 1: p^2	Accept at stage 2: $2p^3q$	Total probability of acceptance
0.9	0.1	0.81	0.1458	0.9558
0.8	0.2	0.64	0.2048	0.8448
0.7	0.3	0.49	0.2058	0.6958
0.6	0.4	0.36	0.1728	0.5328
0.5	0.5	0.25	0.125	0.375
0.4	0.6	0.16	0.0768	0.2368
0.3	0.7	0.09	0.0378	0.1278
0.2	0.8	0.04	0.0128	0.0528
0.1	0.9	0.01	0.0018	0.0118

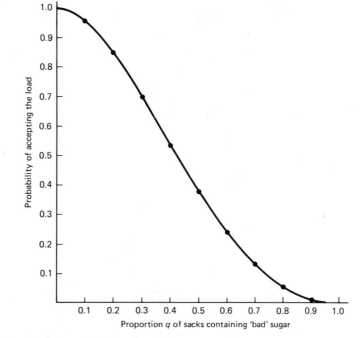

Figure 3.6 Operating characteristic curve for sugar sampling

It may be that the supplier is generally very reliable and rarely sends a bad load. As sugar undergoes a standardised manufacturing process at the refinery, the quality within the load is likely to be fairly uniform. In addition (as mentioned earlier), the viscosity of the mixture can be corrected during the manufacturing process.

Another Two-stage Sampling Scheme

The same confectionery manufacturer also buys coconut for one of their sweet varieties. Coconut is much more variable in quality than sugar and is sampled more rigorously. It has undergone only a simple preparation process (rather than an industrialised refining involving chemicals) and is a relatively natural product, subject to natural variations.

A sample of 10 sacks is taken at stage 1. For the load to be accepted at this stage there must be 9 or 10 good sacks in the sample. If there are 2 or more bad samples, the technicians undertake a stage-2 sampling. 10 more samples are taken and these must all be up to standard for the load to be accepted.

Let p be the proportion of sacks containing good coconut and q the proportion of sacks containing bad coconut. The probability of accepting a load is calculated as either of the following.

1. The load is accepted at the stage 1 (i.e. 9 or 10 samples are satisfactory), i.e. probability $= p^{10} + 10p^9q$.
2. Stage-2 samples are taken, if fewer than 9 of the first batch of samples are acceptable, i.e. probability $= 1 - (p^{10} + 10p^pq)$, *and* all 10 of the stage-2 samples are satisfactory, i.e. probability $= p^{10}$, giving a probability of $[1 - (p^{10} + 10p^9q)]p^{10}$.

The calculations necessary for plotting the operating characteristic curve in Figure 3.7 are given in Table 3.14.

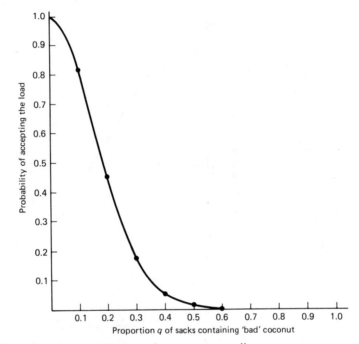

Figure 3.7   Operating characteristic curve for coconut sampling

Table 3.14   Probability calculations

p	q	Accept at stage 1: $p^{10} + 10p^9q$	Accept at stage 2: $p^{10}[1 - (p^{10} + 10p^9q)]$	Total probability of acceptance
0.9	0.1	0.736 1	0.092 0	0.828 1
0.8	0.2	0.375 8	0.067 0	0.442 8
0.7	0.3	0.149 3	0.024 0	0.177 5
0.6	0.4	0.046 3	0.005 7	0.052 0
0.5	0.5	0.010 78	0.000 96	0.011 7
0.4	0.6	0.001 67	0.000 098	0.000 18
0.3	0.7	0.000 146	0.000 006	0.000 02
0.2	0.8	0	0	0
0.1	0.9	0	0	0

The coconut sampling scheme is much more efficient than the sugar sampling scheme, as you can see if you compare the two operating characteristic curves. The coconut sampling curve is much steeper, and the probability of accepting a batch decreases rapidly as the proportion of bad coconut in the load increases.

Section 3.7 **Solutions to Problems**

The Frivolous Pursuit Quiz

The correct answers are as follows.

1. True.
2. True.
3. False (it is in Sudan).
4. True.
5. True.
6. False (he took 13–15 s).
7. True.

Practical Follow-up 3.1: The Examiner's Problem

10 questions are set each with 2 possible alternative answers. The probability that a candidate can gain a pass mark purely by guesswork is 0.623.

Number of correct answers	Number of outcomes	
0	$^{10}C_0 = 1$	
1	$^{10}C_1 = 10$	
2	$^{10}C_2 = 45$	Fail
3	$^{10}C_3 = 120$	
4	$^{10}C_4 = 210$	
5	$^{10}C_5 = 252$	
6	$^{10}C_6 = 210$	
7	$^{10}C_7 = 120$	Pass
8	$^{10}C_8 = 45$	
9	$^{10}C_9 = 10$	
10	$^{10}C_{10} = 1$	

Therefore,

number of outcomes for 5 or more correct = 638

total number of outcomes　　　　　　　 = 1024

Thus,

$$\text{probability of a pass} = \frac{638}{1024}$$

$$= 0.623$$

Section 3.3

In the question about Big Ben, the correct mass is 13.575 tons.

The Examiners Problem Solved(?)

If 10 questions are set each with 4 parts, the probability of getting 5 or more questions correct by guessing is 0.078.

'Frivolous Pursuit'—The Quiz

For each statement write true or false. Please *do not* discuss the questions with your neighbour or copy their answer.

1. When exposed to a wind of 30 mile per hour in a temperature of $-30\,°F$, human flesh freezes solid in 30 s.
2. A coho is a fish.
3. Omdurman is a town in Morocco.
4. The probability of getting a royal flush in poker is about 1 in 650 000 hands.
5. A brool is a deep murmur.
6. Before anaesthetics were invented, the shortest time recorded for a leg amputation was 20 s by Napoleon's chief surgeon, Dominic Larrey.
7. The fonticulus is a little dip just on top of the breast bone.

Chapter 4 The Poisson Distribution

Section 4.1 Random Events

How many cars will arrive at a petrol station in any one minute? Can we predict the pattern of arrivals for the next hour?

How many telephone calls will come into a switchboard in any period of 10 s? How often will the switchboard be 'jammed'?

How many customers are likely to arrive at a baker's shop in any minute? Can the shop cope with this volume of customers?

All these questions involve the consideration of events which seem to occur at random. In this chapter, we observe and study the number of times that a random event occurs in a fixed time or space and look for any patterns in these numbers. If we can devise a mathematical probability model which generates results similar to those we observe, then it can be used to make predictions about future events.

Section 4.2 Practical Data Collection

Practical 4.1 Vehicles Passing an Observation Point

Vehicles are to be counted as they pass along a road. Choose a suitable stretch of road. There must be a smooth flow of traffic, with no traffic lights, roundabouts, sharp bends or other obstructions which would cause the traffic to arrive in bunches. Also avoid obstructions caused by buses and lorries on single-carriageway roads.

We are looking for an example of a random process and need to be able to assume that each vehicle passes independently of every other vehicle. It would not be suitable to take a count during the rush hour; cars will form queues and so not arrive independently.

With this in mind, set up an observation point on a suitable stretch of road. A bridge over a motorway is ideal. You will need a minimum of two people to collect the basic data required. One person acts as time keeper and shouts out 'Now' every 15 s. The recorder counts the number of vehicles passing in each interval of 15 s and logs them on a log sheet. (An example of the type of recording sheet to use is included among the photocopiable pages as page PC8.)

We found that time intervals of 15 s were ideal on a busy motorway but, on a quieter road, 30 s or even 1 min might be more appropriate. If the traffic is dense, you may count only lorries or only cars.

You should aim to collect the data for 75–100 time intervals on one carriageway. If you have extra personnel, you can have another recorder noting the number of vehicles on the opposite carriageway. You may also use this opportunity to collect some data to use in the chapter on the exponential distribution (see Chapter 14 for details). The people doing this should work independently of the others, writing down the time interval between each car. This could be very difficult on a motorway and you might need to concentrate on one lane of one carriageway.

When you have completed your data collection, follow through the data analysis outlined in Sections 4.4 and 4.6.

Section **4.3** **Alternative Practicals**

There are several alternative practicals so that groups of students can work separately. The data collected are all to be analysed in similar way to the example used in this chapter. In this way, the various results can be compared and discussed.

Practical **4.2** **Cars Arriving at a Petrol Station**

Work in a similar way to that described in Practical 4.1 but position yourselves outside a garage (which may be in a town, although preferably not close to traffic lights). Count the number of vehicles arriving to buy petrol in each period of 1 min.

Record your results on the sheet headed **Log Sheet for Collecting Poisson Data** (page PC8).

Practical **4.3** **Customers Arriving at a Shop or Bank**

Practical 4.1 can be modified slightly to observe and record the arrivals of customers at a shop or bank. Time periods of 1 min will probably be appropriate, although in a busy town 30 s may be better. The same data-recording sheet can be adapted for your results.

Practical **4.4** **Telephone Calls**

Arrange to sit in the school general office to log the frequency of telephone calls coming into the switchboard. It would probably be best to count the number of calls in an interval of 5 min. Alternatively, choose an interval to give a mean number of calls in this interval between, say, 3 and 10. Again, as with the car-counting exercise, it will be useful for the unit on the exponential distribution if someone also makes a note of the time intervals between calls. Avoid the early morning 'rush hour' between about 8.30 am and 9.30 am when there may be a rush of telephone calls.

This exercise will take a long time to generate enough data to analyse, but the work could be shared among several students so that a whole day (or half-day) could be covered.

Modify the data sheet page PC8 to record your data.

Practical **4.5** **Radioactive Particles**

This activity, if it is possible, will require help and equipment from the physics department.

Using a Geiger counter to measure the number of particles emitted per minute as general background radiation. You do not need to borrow any radioactive material as a source.

Modify the log sheet in order to record your results.

Practical **4.6** **Weed Count**

Keeping away from paths and hedges, measure out an area on the school playing field 10 m by 10 m (Figure 4.1). Use random numbers in pairs to represent the coordinates of the bottom left-hand corner of your sample.

For example, if the random numbers 1, 5 occur, this is the sample which would be chosen. A square frame made of wood or wire can then be placed on the ground to give a sample of 1 m^2, or you can improvise with four metre-rules.

Take 50 samples. Count the number of dandelions (say), counting only separate plants. Plants which are joined at the roots should be counted as one plant. (You may count another type of weed if you prefer.)

Record your data on a log sheet similar to page PC8.

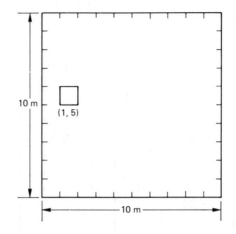

Figure 4.1 The sampling area

Section **4.4** **Looking at the Results**

Here are the results of a survey carried out on the south-bound carriageway of the M1. The number of vehicles passing were counted over intervals of 15 s and gave the frequency distribution in Table 4.1.

Table 4.1 The vehicles on the south-bound carriageway of the M1

Number x of vehicles passing in a 15 s interval	f
0	1
1	4
2	8
3	6
4	17
5	18
6	13
7	3
8	3
9	1
10	1
11	1
12	0

This is the sort of distribution that you can expect provided that there are no road conditions to cause bunching of traffic.

These results are illustrated by the line graph in Figure 4.2.

Looking at the general shape of the diagram the frequencies for 3 vehicles and for 7 vehicles appear to be too low, but the general shape of the distribution (apart from these values) is close to what we might expect. Since the sample is fairly small, we expect relatively more variation from the expected values.

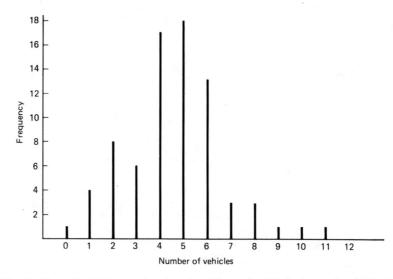

Figure 4.2 Number of vehicles passing junction 33 on the M1 in intervals of 15 s (south-bound carriageway)

Next we calculate the mean and variance in each time interval of 15 s (Table 4.2).

$$\bar{x} = 4.59 \approx 4.6$$

$$\text{variance} = \frac{\sum fx^2}{\sum f} - \left(\frac{\sum fx}{\sum f}\right)^2$$

$$= \frac{1921}{76} - \left(\frac{349}{76}\right)^2$$

$$= 25.27 - (4.59)^2$$

$$= 25.27 - 21.08$$

$$= 4.19$$

Note that the mean and the variance are close in value.

Table 4.2 Data for calculation of the mean and variance

x	f	fx	fx^2
0	1	0	0
1	4	4	4
2	8	16	32
3	6	18	54
4	17	68	272
5	18	90	450
6	13	78	468
7	3	21	147
8	3	24	192
9	1	9	81
10	1	10	100
11	1	11	121
Total	76	349	1921

Practical Follow-up **4.1**

1. Tabulate your results into a frequency distribution similar to ours in Table 4.1.
2. Draw a line graph to illustrate your frequency distribution.
3. Calculate the mean and variance of your data.

Section **4.5** **Finding a Probability Model**

There are several ways that we can look at this situation. In our experiment, we observed a total of 349 vehicles over a period of 76 time intervals of 15 s.

Now, supposing that we knew from the beginning that these 349 vehicles were going to pass by, that they would pass by independently and that they were equally likely to arrive in any one of these time intervals of 15 s.

Consider any one of these time intervals. The probability that the first car arrives in it is 1/76. Since vehicles arrive independently, the probability that the second arrives in this interval is also 1/76 and so it is for each of the 349 vehicles. Hence the number of cars arriving in this time period will have a binomial distribution with $n = 349$, $p = 1/76$, i.e. $B(349, 1/76)$ (see Chapter 3 on the binomial distribution).

Now, unfortunately, we do *not* know beforehand how many vehicles are going to arrive. All that we are likely to know from experience is the mean number of vehicles per interval of 15 s. We do notice, however, that, the longer we make the observations, the larger the number of vehicles and the smaller the probability of a particular vehicle arriving in a given time period, i.e. the larger is n and the smaller is p from $B(n, p)$.

Our theoretical distribution is therefore that which the $B(n, p)$ becomes as n increases and p decreases in such a way that the mean, np, remains a constant.

This distribution is called the Poisson distribution with mean μ and we need to find the probability function in terms of μ.

Now, if $X \sim B(n, p)$, we have

$$P(x = k) = {}^nC_k p^k q^{n-k} = \frac{n!}{(n-k)!k!} p^k q^{n-k}$$

$$P(x = k+1) = {}^nC_{k+1} p^{k+1} q^{n-k-1} = \frac{n!}{(n-k-1)!(k+1)!} p^{k+1} q^{n-k-1}$$

so that

$$P(x = k+1) = P(x = k)\frac{(n-k)p}{(k+1)q}$$

$$= P(x = k)\frac{np - kp}{(k+1)(1-p)}$$

However, $np = \mu$ and therefore

$$P(x = k+1) = P(x = k)\frac{\mu - kp}{(k+1)(1-p)}$$

Now let $p \to 0$ to get the distribution of the limiting variable y:

$$P(y = k+1) = \frac{\mu}{k+1} P(y = k)$$

So, for this limiting (Poisson) distribution,

$$P(Y=1) = \frac{\mu}{1} P(Y=0)$$

$$P(Y=2) = \frac{\mu}{2} P(Y=1)$$

$$= \frac{\mu^2}{2!} P(Y=0)$$

and similarly

$$P(Y=r) = \frac{\mu^r}{r!} P(Y=0) \qquad r = 0, 1, 2, \ldots, \infty$$

Now, since the sum of the probabilities is 1,

$$1 = \sum_{r=0}^{\infty} P(Y=r)$$

$$= P(Y=0)\left(1 + \mu + \frac{\mu^2}{2!} + \frac{\mu^3}{3!} + \ldots + \frac{\mu^r}{r!} + \ldots \right)$$

$$= P(Y=0)e^{\mu}$$

Hence

$$P(Y=0) = e^{-\mu}$$

and the Poisson distribution has the probability function

$$P(Y=r) = e^{-\mu}\frac{\mu^r}{r!} \qquad r = 0, 1, 2, \ldots, \infty$$

$$= 0 \qquad\qquad \text{otherwise}$$

Another way to look at the process is as follows.

1. The vehicles arrive *randomly* over time.
2. At any given instant of time there can be at most one vehicle passing the observation point, i.e. they arrive *singly*. (This is why we only look at the vehicles in one carriageway.)
3. The vehicles arrive in such a way that the expected number arriving in a given time interval of 30 s is twice that arriving in an interval of 15 s. Generally, if vehicles are arriving at a rate of λ per second, the expected number of vehicles arriving in a time period of t is λt. In this sense the cars are said to arrive *uniformly*, and we assume that this average rate of arrivals is fairly constant from day to day at that time.
4. The cars arrive *independently*, i.e. the probability of a vehicle arriving in any small time interval is independent of a vehicle arriving in any other small interval.
5. The variable that we are studying is the number of events (vehicles arriving) in an interval of a given size.

On the understanding that intervals may be over space rather than time, the above five assumptions define what is called a *Poisson process* and the distribution is called a Poisson distribution.

The same formula for the probability function can be deduced directly from the above axioms but, since the proof involves the solution of differential equations, it is not reproduced here.

An alternative approach is to imagine our time interval of 15 s divided into n small time intervals of δt. These small intervals must be so short that there cannot be more than one vehicle arriving in δt. (It follows then that n will be a large number.) From this idea, we could give a meaning to the probability p that a car arrives in a time interval δt:

$$p(\text{vehicle arriving in } \delta t) = \frac{\text{mean number of vehicles in 15 s}}{\text{number of intervals } \delta t \text{ in 15 s}}$$

$$= \frac{\mu}{n}$$

Hence,

$$p(\text{no vehicle arrives in } \delta t) = 1 - \frac{\mu}{n}$$

The corresponding binomial probability of r vehicles in the 15 s is given by

$$P(X = r) = {}^{n}C_{r}p^{r}q^{n-r}$$

$$P(r) = \binom{n}{r}\left(\frac{\mu}{n}\right)^{r}\left(1 - \frac{\mu}{n}\right)^{n-r}$$

We have specified that δt must be very small; so $\delta t \to 0$ and hence $n \to \infty$.

The limit of this expression is

$$P(r) = \frac{\mu^{r} e^{-\mu}}{r!}$$

Section 4.6   Calculation of Poisson Probabilities

For our vehicle data, the mean is 4.6. We can investigate whether the observed frequencies are close to those from a Poisson distribution, but which Poisson distribution? The values depend only on the mean of the distribution; so it seems sensible to compare the data with the Poisson distribution with mean 4.6.

The general term for a Poisson distribution as in Table 4.3 is

$$\frac{e^{-\mu}\mu^{y}}{y!}$$

For calculation purposes, it is convenient to use the result that

$$P(Y = k + 1) = \frac{\mu}{k + 1}P(Y = k)$$

provided that care is taken not to build up errors from a small value of $P(y = 0)$ which has rounding errors.

So, for example,

$$P(Y = 5) = P(Y = 4)\frac{4.6}{5}$$

Table 4.3 Probability functions for a Poisson distribution

Probability function	Value	
$P(Y=0)$	$e^{-4.6}$	≈ 0.01005
$P(Y=1)$	$e^{-4.6} \times 4.6$	≈ 0.0462
$P(Y=2)$	$e^{-4.6} \times \dfrac{(4.6)^2}{2!}$	≈ 0.1063
$P(Y=3)$	$e^{-4.6} \times \dfrac{(4.6)^3}{3!}$	≈ 0.1631
$P(Y=4)$	$e^{-4.6} \times \dfrac{(4.6)^4}{4!}$	≈ 0.1875
$P(Y=5)$	$e^{-4.6} \times \dfrac{(4.6)^5}{5!}$	≈ 0.1725
$P(Y=6)$	$e^{-4.6} \times \dfrac{(4.6)^6}{6!}$	≈ 0.1323
$P(Y=7)$	$e^{-4.6} \times \dfrac{(4.6)^7}{7!}$	≈ 0.0869
$P(Y=8)$	$e^{-4.6} \times \dfrac{(4.6)^8}{8!}$	≈ 0.0500
$P(Y=9)$	$e^{-4.6} \times \dfrac{(4.6)^9}{9!}$	≈ 0.0255
$P(Y=10)$	$e^{-4.6} \times \dfrac{(4.6)^{10}}{10!}$	≈ 0.0118
$P(Y=11)$	$e^{-4.6} \times \dfrac{(4.6)^{11}}{11!}$	≈ 0.0049

$$= 0.1875 \times \frac{4.6}{5}$$

$$= 0.1725 \text{ as above}$$

To calculate the expected frequencies, multiply each Poisson probability by the total frequency which was 76 (Table 4.4 and Figure 4.3).

Practical Follow-up 4.2

1. Calculate the Poisson probabilities for a distribution with the mean calculated from your own practical results.
2. Find the expected frequencies predicted by the Poisson model by multiplying each probability by the total number of observations in your sample.
3. Compare your observed frequency distribution with these theoretical expected frequencies. Does the Poisson model produce predictions which are close to your observed results?

A Further Example: The M1 North-bound Carriageway

Table 4.5 shows the results of a survey carried out simultaneously in the same intervals of 15 s on the north-bound carriageway of the M1.

Table 4.4 Calculation of expected frequency

y	$p(y)$	$76p(y)$	Expected frequency to nearest whole number
0	0.0101	0.76	1
1	0.0462	3.51	4
2	0.1063	8.08	8
3	0.1631	12.40	12
4	0.1875	14.25	14
5	0.1725	13.11	13
6	0.1323	10.05	10
7	0.0869	6.60	7
8	0.0500	3.8	4
9	0.0255	1.9	2
10	0.0118	0.89	1
11	0.0049	0.37	0

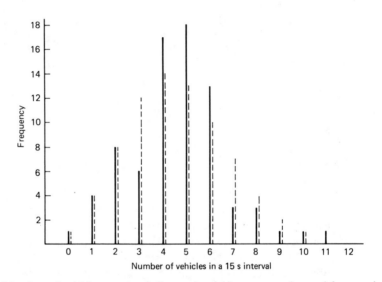

Figure 4.3 Number of vehicles passing in intervals of 15 s: ———, observed frequencies from M1 survey; – – –, theoretical expected Poisson frequencies for $\mu = 4.6$

Calculate the mean and variance of the distribution. What do you notice? Calculate the Poisson probabilities for a distribution of that mean and then find the expected frequencies.

Draw the observed distribution and the expected frequencies on the same line diagram.

Table 4.5 Number of vehicles passing in a 15 s interval

Number of vehicles passing in a 15 s interval	0	1	2	3	4	5	6	7	8	9	10
Frequency	0	6	5	11	15	10	6	11	7	2	1

Section 4.7   The Sum of Two Poisson Variables

By referring to the original log sheets used when you collected your data, you may now investigate what happens when two Poisson variables are summed together.

Intervals of 30 s on the M1 South-bound Carriageway

By taking the sum of two intervals of 15 s, you obtain the number of vehicles passing in an interval of 30 s. The 76 observations for intervals of 15 s become 38 observations for intervals of 30 s from the data that we collected. Table 4.6 shows the results.

Table 4.6   Number of vehicles passing in a 30 s interval

Number of vehicles passing in a 30 s interval	Frequency
3	0
4	1
5	1
6	2
7	6
8	3
9	7
10	7
11	6
12	3
13	1
14	1

The mean of these data is 9.18. The mean for our 15 s intervals was 4.59. We might expect our data for 30 s intervals also to follow a Poisson distribution, and the mean obtained for the 30 s intervals to be equal to twice the mean for the 15 s intervals.

If the Poisson probabilities are calculated for $\mu = 9.2$, the expected frequencies can be obtained. Figure 4.4 is a diagram showing the observed and expected frequencies.

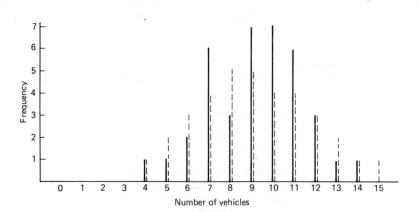

Figure 4.4   Number of vehicles passing in intervals of 30 s (south-bound carriageway): ———, observed frequencies; –––, expected Poisson frequencies for $\mu = 9.18$

Distribution of Vehicles Passing North and South Simultaneously

Because of the way that we collected our data, we were able to add together the results for north-bound and south-bound traffic from the original log sheets. This gives a distribution of the total amount of traffic passing the observation point in a 15 s interval (Table 4.7).

Table 4.7   Total traffic in a 15 s interval

Total traffic north and south in a 15 s interval	Frequency
3	1
4	1
5	5
6	4
7	5
8	14
9	9
10	10
11	5
12	9
13	8
14	0
15	1
16	2

The mean for this distribution is 9.43. The means for south-bound and north-bound traffic were 4.59 and 4.81 respectively, giving a total of 9.4.

The Poisson probabilities and expected frequencies for this distribution are shown in Figure 4.5.

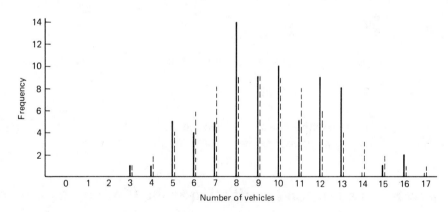

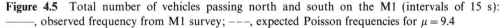

Figure 4.5   Total number of vehicles passing north and south on the M1 (intervals of 15 s): ———, observed frequency from M1 survey; – – –, expected Poisson frequencies for $\mu = 9.4$

The Sum of Two Poisson Random Variables

Vehicles are being counted separately on north-bound and south-bound carriageways with respective means of α and β per 15 s interval. If r is the total number of vehicles passing in both directions, what will the distribution of r be?

Firstly, we must assume that the two flows of traffic are independent of each other (and this seems a reasonable assumption). Thus the probability of observing n vehicles which are north bound and $r - n$ which are south bound is

$$\frac{e^{-\alpha} \alpha^n}{n!} \frac{e^{-\beta} \beta^{(r-n)}}{(r-n)!}$$

In order that the total will be r, we could have

    0 which are north bound and r which are south bound

or

    1 which is north bound and $r - 1$ which are south bound

or

    2 which are north bound and $r - 2$ which are south bound

$$\vdots \qquad\qquad\qquad\qquad\qquad \vdots$$

or

$r - 1$ which are north bound and   1 which are south bound

or

r which are north bound and   0 which are south bound

The total probability of observing r vehicles is the sum of the probabilities of all these ways of splitting up r between the two types:

$$P(r) = \sum_{n=0}^{r} \frac{e^{-\alpha} \alpha^n}{n!} \frac{e^{-\beta} \beta^{(r-n)}}{(r-n)!}$$

$r!/r!$ is introduced as an extra term:

$$P(r) = \frac{e^{-(\alpha+\beta)}}{r!} \sum_{n=0}^{r} \frac{r!}{n!(r-n)!} \alpha^n \beta^{(r-n)}$$

The term inside the summation is simply the binomial expansion of $(\alpha + \beta)^r$:

$$P(r) = \frac{e^{-(\alpha+\beta)}}{r!} (\alpha + \beta)^r$$

This shows that the sum of two independent Poisson random variables with means α and β forms another Poisson random variable with mean $\alpha + \beta$.

Section 4.8   The Poisson Distribution as the Limit to the Binomial Distribution

In this experiment, we take samples from a population where a binomial distribution should really apply, but where the probability p of a success is very small. Also, n should be very large but this may be difficult in a practical situation but, provided that we are only interested in probabilities to two or three decimal places, this should not be problematical. Some examples are given later. The materials needed for the experiment can be adapted according to what is available.

Practical **4.7   Poisson Distribution as the Limit to the Binomial Distribution**

In a large box or bowl, place any of the following populations from which samples are to be taken as described. Each sample is replaced after the result has been recorded, and the population shaken well. Take 100 samples.

Choose from one of the following.

1. A 500 g packet of red lentils into which a tablespoonful of green lentils has been stirred well.
2. Pieces of card or paper of two colours mixed in the proportions 20 to 1.
3. Waste from a hole puncher or paper tape machine could be used provided that about 1/20 of the quantity is of a different colour.

For population 1, take a sample of size 900. (This is about a dessertspoonful, but the same number must be taken each time.) For populations 2 or 3, take samples of 20 and record the number of the rare colour in each group.

Results

Table 4.8 shows some results obtained using a quantity of green lentils mixed with red lentils.

Table 4.8   Number of green lentils in a mix of green and red lentils

Number of green lentils	Frequency
0	0
1	0
2	6
3	10
4	20
5	19
6	14
7	11
8	11
9	10
10	2
11 or more	0

The mean and the variance for these data are 5.53 and 5.09, respectively. The line graph below shows the observed frequencies and the expected frequencies from a Poisson distribution with the same mean (Fig. 4.6).

The differences between the observed frequencies and expected frequencies are not very great for this small sample. In Chapter 13 (on the χ^2 test), there is a technique which will show that these differences are not statistically significant. The same is true for all the examples in this chapter.

The experiments described in this section can all be taken as examples which can be modelled by the binomial distribution. Each item can be categorised into one of two groups, success and failure, and it is possible to assign values for p and n.

The results obtained look like Poisson data and yet the samples have originated from binomial sampling. In both this last experiment and the observations that you made at the beginning of the unit, you should have found that the mean and variance were close in value so that

mean $\approx$ variance

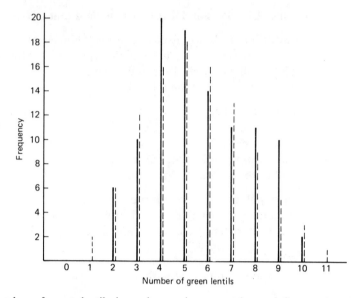

Figure 4.6 Number of green lentils in each sample: ——, observed frequencies; ---, expected Poisson frequencies for $\mu = 5.53$

i.e.

$$np \approx np(1 - p)$$

From this it follows that $1 - p \approx 1$, i.e. the probability of a failure is nearly 1 (and the probability of a success very small). As the sample size n increases, the Poisson distribution turns out to be a useful approximation for the binomial distribution. In these situations where n is large, it is very difficult to calculate binomial probabilities in any case, and computation is much easier using the Poisson approximation.

A proof was given in Section 4.5 that the Poisson distribution is the limit of the binomial distribution as $n \to \infty$ with

mean $= np$

$$\mu = np$$

so that

$$p = \frac{\mu}{n}$$

and

$$1 - p = 1 - \frac{\mu}{n}$$

Comparison of Binomial and Poisson Probabilities

Table 4.9 shows the probabilities for $n = 30$, $p = 0.1$ calculated using both a binomial distribution and a Poisson distribution.

With such a low value of p the resulting distribution is heavily weighted towards the low values of x. The probabilities for $x > 12$ are less than 0.00001. (This distribution is said to be positively skewed.)

Table 4.9 Binomial and Poisson probabilities for $n = 30$, $p = 0.1$

x	Binomial probability $B(30, 0.1)$	Poisson probability $\mu = 3$
0	0.0424	0.0498
1	0.1413	0.1494
2	0.2277	0.2240
3	0.2361	0.2240
4	0.1171	0.1680
5	0.1023	0.1008
6	0.0474	0.0504
7	0.0180	0.0216
8	0.0058	0.0081
9	0.0016	0.0027
10	0.0004	0.0008
11	0.0001	0.0002
12	0.0000	0.0001

Note that the Poisson approximation is correct to the first decimal place and within 2 for the second decimal place.

If p is decreased by a factor of 10 to 0.01, the correspondence between the two sets of probabilities improves (Table 4.10). We have agreement to three decimal places (but note how quickly the probabilities 'disappear' beyond the fourth decimal place, for $x > 4$).

Table 4.10 Binomial and Poisson probabilities for $n = 30$, $p = 0.01$

x	Binomial probability $B(30, 0.01)$	Poisson probability $\mu = 0.3$
0	0.7397	0.7408
1	0.2242	0.2222
2	0.0328	0.0333
3	0.0031	0.0033
4	0.0002	0.0003
5	0.0000	0.0000
6	0.0000	0.0000
7	0.0000	0.0000
8	0.0000	0.0000
9	0.0000	0.0000

Proof that the Poisson Distribution is the Limit of the Binomial Distribution

Using Generating Functions

For the binomial distribution the probability-generating function is

$$E(t^x) = \sum_0^n \binom{n}{x} p^x (1 - p)^{n-x} t^x = (1 - p + pt)^n$$

and the moment-generating function is

$$E(e^{tx}) = \sum_{0}^{n} \binom{n}{x} p^x (1-p)^{n-x} e^{tx} = (1 - p + p e^t)^n$$

For the Poisson distribution the probability-generating function is

$$E(t^x) = \sum_{0}^{\infty} e^{-\mu} \frac{\mu^x}{x!} t^x = e^{-\mu} e^{\mu t} = e^{\mu(t-1)}$$

and the moment-generating function is

$$E(e^{tx}) = \sum_{0}^{\infty} e^{-\mu} \frac{\mu^x}{x!} e^{tx} = e^{-\mu} e^{\mu e^t} = e^{\mu(e^t - 1)}$$

We use the result that

$$\lim_{n \to \infty} \left(1 + \frac{a}{n}\right)^n = e^a$$

Therefore the probability-generating function and the moment-generating function for the binomial distribution are $(1 - p + pt)^n$ and $(1 - p + pe^t)^n$, respectively, i.e. $[1 + (\mu/n)(t-1)]^n$ and $[1 + (\mu/n)(e^t - 1)]^n$, respectively. Let $n \to \infty$; then we get $e^{\mu(t-1)}$ and $e^{\mu(e^t - 1)}$ which are the probability-generating function and the moment-generating function, respectively, of the Poisson distribution.

Hence the limit as $n \to \infty$ with $np = \mu$ of the binomial distribution is the Poisson distribution.

Section 4.9 Solution to Problem

A Further Example: The M1 North-bound Carriageway

The answers are as follows:

$\bar{x} = 4.81$

sample variance $= 4.97$

x	Poisson probabilities	Expected frequencies
0	0.0082	0
1	0.0395	2.9
2	0.0948	7
3	0.1517	11.2
4	0.1820	13.5
5	0.1747	12.9
6	0.1398	10.3
7	0.0959	7.1
8	0.0575	4.25
9	0.0307	2
10	0.0147	1.08

Chapter 5 The Normal Distribution

Section 5.1   Poisonous or Edible?

Collectors of fungi have to be very careful to identify species correctly, as in some cases poisonous varieties may be very close in appearance to edible ones. The edible *Hypholoma capnoides* is a pleasant-tasting mushroom which grows abundantly in coniferous forests. Unfortunately, it looks very much like the Sulphur Tuft which is poisonous. There are slight differences in colour, the Sulphur Tuft being more yellow, but another more important difference is size.

Figure 5.1 shows the average size of mature plants with the Sulphur Tuft growing much larger than the *Hypholoma capnoides*. However, some of the edible mushrooms will grow larger than average, and likewise some of the Sulphur Tufts will not grow as tall as the average for that species. So there may well be some overlap between the sizes of edible and poisonous varieties. What we require is some idea of the distribution of the heights and sizes of each type, i.e. a population model for each species.

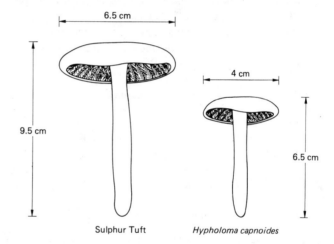

Figure 5.1   A comparison of the sizes of the Sulphur Tuft and *Hypholoma capnoides* (Klan, 1981)

There are many other examples among plants, insects and animals where closely related species can be distinguished by size. A difficult problem for ornithologists is the identification of Little Buntings (average length, 13.5 cm) and Reed Buntings (average length, 15 cm). The appearance of these two birds is very similar and their size is a helpful clue. Again, however, we shall need more precise information, in terms of population models for the distributions of sizes. As ornithologist D. Wallace (1980) writes, '...while female Littles are always smaller than female Reeds, some large male Littles overlap, at least in wing length, with them.'

Section 5.2   Practical Data Collection

The following suggestions for data collection do not include picking fungi in case of misidentification! However, most naturally occurring populations form distributions of a similar shape; so you may choose any of the ideas outlined in this section.

In this chapter, we are concerned with the distribution of measurements (such as heights, masses, etc.) of large populations. It will not be possible for you to collect large amounts of data in a limited time, but even so you should aim to measure a sample of at least 100. A sample of even this size will sometimes yield an atypical distribution, especially if it is not a true random sample.

Again, because of time constraints, we have not drawn up a scheme for random sampling in this unit. However, if you are thinking of extending some of the ideas into a larger project, you should consider carefully your sampling method.

Practical 5.1   Leaf Lengths and Widths

This is a survey which can be completed very quickly by a group of students sharing the task. Take a sample of 100 leaves of the same species, measure and record the length and width of each leaf. If this is done, the data can be used for a later chapter on correlation and regression. Take measurements to the nearest millimetre.

By looking at the frequency distribution of leaf lengths (or widths) in the sample, we can devise a suitable mathematical model for the population of such leaves.

Here we are dealing with continuous data although the measurements recorded will involve some compromise of accuracy according to the measuring instrument used. If an ordinary ruler is used and measurements taken to the nearest millimetre, the results can be inaccurate to 0.5 mm either way.

Practical 5.2   Heights and Weights

Measure the heights and weights of 50 people. If you cannot obtain a group of adults, ensure that you use children all the same sex and age group. Record your raw data with the height and weight of each person together, as this information will be very useful for a later chapter. Measurements should be taken using centimetres and kilograms to one decimal place of accuracy.

Practical 5.3   Biology or Rural Studies

If you are studying biology or rural studies, you may have access to a large number of plants or insects of the same species and age, which you can measure in height or length.

Practical 5.4   Distances

Measure the distance travelled down a slope by a toy car. You will need to construct a slope similar to this, although the actual dimensions are not important (Figure 5.2). If you have a Scalextric set, you may be able to adapt this idea, using a longer section of track. A long strip of graph paper can be Sellotaped to the track and used as a measuring scale.

Place a small toy car at the top of the slope and let go. (Do not push it.) The distances travelled will form a distribution, which you can examine to see whether it is normal.

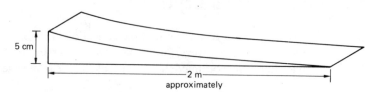

Figure 5.2 Constructed slope

Practical **5.5**   **Coins I**

Weigh separately 50–100 coins of the same denomination (2 p coins are suitable). In order to do this, and to detect the differences in weights, you will need a very sensitive balance.

Practical **5.6**   **Coins II**

Obtain permission from a bank to weigh a large sample (100 +) of £1 coins singly. If you were designing a vending machine operated by £1 coins, what limits would you set on the machine? Bear in mind that you do not want to allow fake or foreign coins such as 100 peseta coins to operate the machine.

Section 5.3   Some Typical Results

When the raw data have been collected, draw up a frequency table for the measurements you have taken grouped into about 10 or 12 groups. If too many groups are used, the data will be too fragmented and the underlying distribution will not show up. If you have lengths and widths of leaves, or heights and weights of people, for example, keep the two distributions separate. Analyse your results as shown in the following example.

These measurements were taken from 100 laurel leaves. The frequency tables (Tables 5.1 and 5.2) and histograms (Figures 5.3 and 5.4) show that both the lengths and the widths form distributions which are similar in shape, with high frequencies at the centre of the range.

Table 5.1   Lengths of 100 laurel leaves

Length (mm)	Frequency
80–89	1
90–99	0
100–109	3
110–119	10
120–129	19
130–139	21
140–149	23
150–159	11
160–169	8
170–179	1
180–189	2
190–199	1

Table 5.2   Widths of 100 laurel leaves

Width (mm)	Frequency
30–34	3
35–39	14
40–44	29
45–49	28
50–54	15
55–59	5
60–64	5
65–69	1

If we had taken larger samples, say of 1000 or 10 000 leaves, we would have been able to use smaller class intervals for classifying the data. In that case the resulting histogram would become smoother as the sample size increased, and so it is more obvious that the underlying distribution could be modelled by a curve.

The distribution of lengths for the entire population of laurel leaves might look like Figure 5.5.

Similarly the distribution of widths might look like Figure 5.6.

Both the lengths and the widths of leaves give distributions of similar shapes. We shall see later that human heights also form distributions of this shape (males and females must be considered separately). These bell-shaped distributions are modelled by different examples of the *normal* distribution.

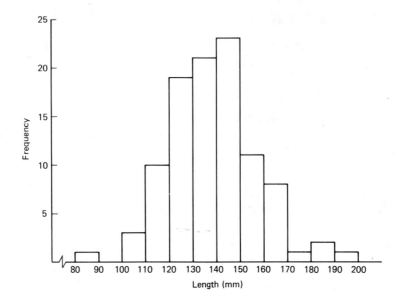

Figure 5.3   The lengths of 100 laurel leaves

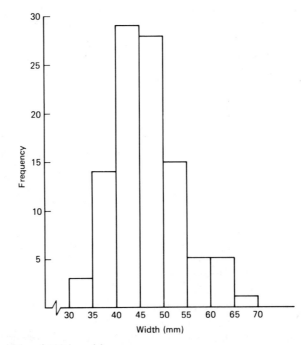

Figure 5.4   The widths of 100 laurel leaves

If the mean and standard deviations are calculated, the raw data can be reclassified in terms of standard deviations from the mean. For example, the mean length for the leaves that we used was 138.1 mm and the standard deviation 18.4 mm. When the raw measurements are grouped in class intervals of 18.4 mm, working each way from the mean we get Table 5.3.

This shows how many leaves are obtained within multiples of the standard deviation

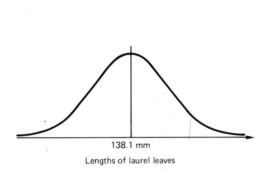

Lengths of laurel leaves

Figure 5.5 Distribution of lengths of laurel leaves

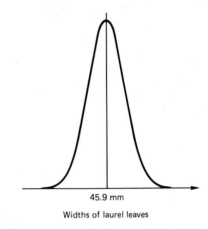

Widths of laurel leaves

Figure 5.6 Distribution of widths of laurel leaves

from the mean. If we do the same for the widths of the leaves, we can compare the distribution again in terms of units of standard deviations.

The data for the width of leaves gave the results in Table 5.4, when the raw data were reclassified in units of the standard deviation (7.1 mm) away from the mean (45.9 mm).

When the two distributions are now drawn, the similarities become even more apparent. Note that the bulk of the data lie within two standard deviations of the mean, and there are no values more than three standard deviations from the mean.

Standardised Frequency Distributions

It appears that, if we take normal distributions with different means and standard deviations and measure in units of the standard deviation above and below the mean, we get a standard distribution (Figures 5.7 and 5.8).

Table 5.3 Reclassification of raw data on the lengths in terms of standard observations from the mean

	Lengths of laurel leaves (mm)	Frequency
Mean − 3 standard deviations		
	82.90–101.2	1
Mean − 2 standard deviations		
	101.3–119.6	12
Mean − 1 standard deviation		
	119.7–138.0	36
Mean		
	138.1–156.4	37
Mean + 1 standard deviation		
	156.5–174.8	11
Mean + 2 standard deviations		
	174.9–193.2	3
Mean + 3 standard deviations		

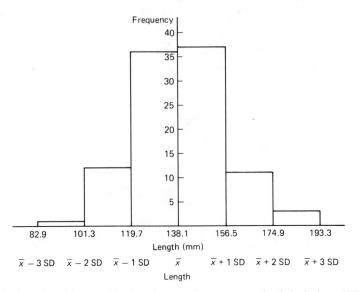

Figure 5.7 The lengths of leaves. The lengths are shown as standard deviations (SDs) away from the mean

The Normal Curve

The normal distribution which models the data that we have obtained will have a mean μ and variance σ^2. Its shape is shown in Figure 5.9.

If we measure in units of σ away from the mean we get a curve of the same shape but labelled as shown below. This is called the standard normal curve. It has a mean 0 and a variance 1. It is obtained by taking

$$Z = \frac{X - \mu}{\sigma}$$

i.e. if $X \sim N(\mu, \sigma^2)$ and $Z = (X - \mu)/\sigma$ then $Z \sim N(0, 1)$ (Figure 5.10).

Table 5.4 Reclassification of raw data on the widths in terms of standard deviations from the mean

	Widths of laurel leaves (mm)	Frequency
Mean − 3 standard deviations		
	24.6–31.6	1
Mean − 2 standard deviations		
	31.7–38.7	12
Mean − 1 standard deviation		
	38.8–45.8	40
Mean		
	45.9–52.9	33
Mean + 1 standard deviation		
	53.0–60.0	9
Mean + 2 standard deviations		
	60.1–67.1	5
Mean + 3 standard deviations		

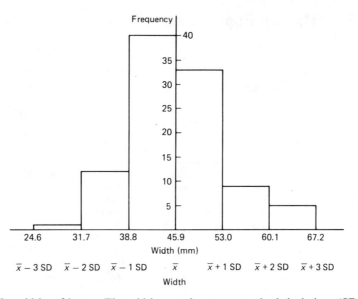

Figure 5.8   The widths of leaves. The widths are shown as standard deviations (SDs) away from the mean

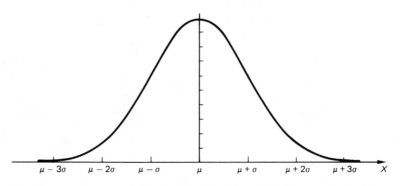

Figure 5.9   The shape of the normal distribution $X \sim N(\mu, \sigma^2)$ which models the data

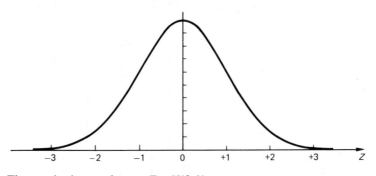

Figure 5.10   The standard normal curve $Z \sim N(0, 1)$

The normal distribution is symmetrical about the mean. Over 95% of the distribution occurs within two standard deviations of the mean either side. The point of inflexion in the curve occurs at $X = \pm \sigma$ (or $Z = \pm 1$) (one standard deviation each side of the mean).

Section 5.4   **Human Populations**

A survey of 10 000 adults conducted in 1980 by the Office of Population Censuses and Surveys (OPCS) found the distributions of heights for males and females (in Great Britain) given in Table 5.5.

The corresponding distributions have been drawn together in Figure 5.11, and you will

Table 5.5   All adults aged 16–64 years (reproduced by permission of the OPCS)

Females		Males	
Height (cm)	All females (%)	Height (cm)	All males (%)
Up to 150	4	Up to 160	2
150.1–152.5	5	160.1–162.5	2
152.6–155.0	8	162.6–165.0	5
155.1–157.5	13	165.1–167.5	8
157.6–160.0	16	167.6–170.0	12
160.1–162.5	15	170.1–172.5	13
162.6–165.0	15	172.6–175.0	14
165.1–167.5	10	175.1–177.5	14
167.6–170.0	7	177.6–180.0	12
170.1–172.5	4	180.1–182.5	7
172.6–175.0	2	182.6–185.0	5
175.1–177.5	1	185.1–187.5	3
Over 177.5	0	187.6–190.0	1
		190.1–193.5	1
		Over 193.5	0
Average value	160.9	Average value	173.8

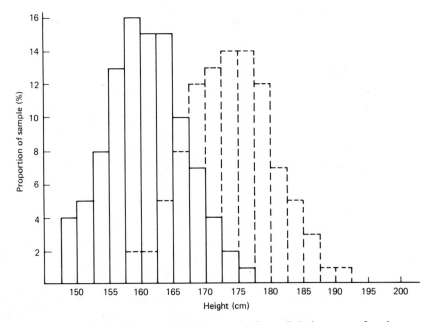

Figure 5.11   Heights of all adults aged 16–64 years in Great Britain: ———, females; – – –, males

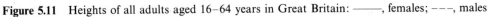

see that both have the same sort of normal distribution shape. The female distribution is to the left of the male distribution as their mean height is lower.

If you have conducted your own survey for heights, you might be interested in the distributions for 16–19 year olds given in the same OPCS publication (Table 5.6). Note that in both cases (males and females) the mean heights are higher than for the whole age range 16–64 years.

Why do you think that this should be so?

Table 5.6 Heights for 16–19 year olds in Great Britain (reproduced by permission of the OPCS)

Females		Males	
Heights (cm)	All females (%)	Height (cm)	All males (%)
		Up to 160	2
		160.1–162.5	2
Up to 150	2	162.6–165.0	4
150.1–152.5	4	165.1–167.5	5
152.6–155.0	6	167.6–170.0	10
155.1–157.5	11	170.1–172.5	15
157.6–160.0	19	172.6–175.0	13
160.1–162.5	16	175.1–177.5	17
162.6–165.0	15	177.6–180.0	14
165.1–167.5	11	180.1–182.5	8
167.6–170.0	8	182.6–185.0	5
170.1–172.5	5	185.1–187.5	2
172.6–175.0	2	187.6–190.0	2
175.1–177.5	1	190.1–192.5	0
Over 177.5	0	Over 192.5	1
Average value	161.7	Average value	174.6

Section 5.5   The Formula for the Normal Distribution

The normal curve was studied by the German mathematician Gauss (1777–1855) and by the Frenchman Laplace (1749–1842). It is sometimes called the Gaussian curve, but most textbooks refer to it as the normal curve.

We have seen already that normal distributions with different means and variances can be standardised by taking the mean as the origin, and by working with the standard deviation as the unit of measurement on the x axis. If this is done, we arrive at the standard normal curve with the form shown in Figure 5.12.

A probability density function for $N(0, 1)$ must have the following properties.

Property 1 $f(z) \geqslant 0$ for all z.
Property 2 The area under the curve is 1, i.e.

$$\int_{-\infty}^{\infty} f(z)\, dz = 1$$

These properties are true of all probability density functions.

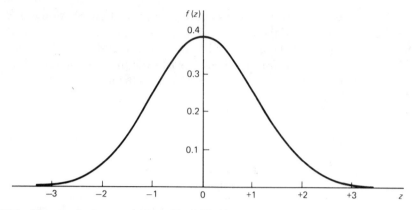

Figure 5.12   The standard normal curve $Z \sim N(0, 1)$

Property 3 $f(-z) = f(z)$ as the curve is symmetrical about $z = 0$.
Property 4 $f(z) \to 0$ as $z \to \pm \infty$ (i.e. the z axis is an asymptote at extreme values of z).
Property 5 $f(z)$ has a maximum at $z = 0$.
Property 6   The mean equals 0:

$$\int_{-\infty}^{\infty} zf(z)\, dz = 0$$

Property 7   The variance equals 1:

$$\int_{-\infty}^{\infty} z^2 f(z)\, dz = 1$$

One function which has these properties is

$$\frac{1}{\sqrt{2\pi}} e^{-z^2/2}$$

Property 1 $f(z) \geq 0$ since $e^y \geq 0$ for all y.
Property 2   A proof is indicated below. The constant factor ensures that the area is 1.
Property 3   Since it is a function of z^2, $f(z) = f(-z)$ and the curve is symmetrical about the axis $z = 0$.
Property 4   Since $e^{-z^2} \to 0$ as $z \to \pm \infty$, this property holds.
Property 5   Since

$$e^{-z^2/2} = \frac{1}{e^{z^2/2}}$$

and so is a maximum when $e^{z^2/2}$ is a minimum; this will be when z^2 is a minimum, i.e. $z = 0$.

Property 6 $\mu = \dfrac{1}{\sqrt{2\pi}} \displaystyle\int_{-\infty}^{\infty} z e^{-z^2/2}\, dz$

Since $g(z) = z e^{-z^2/2}$ is an odd function of z, i.e. $g(z) = -g(-z)$, for all z this is zero.

Property 7   Integrating by parts, we find that

$$\frac{1}{\sqrt{2\pi}} \int_{-\infty}^{\infty} z^2 e^{-z^2/2} \, dz = \frac{1}{\sqrt{2\pi}} \left(\left[-z e^{-z^2/2} \right]_{-\infty}^{\infty} + \int_{-\infty}^{\infty} e^{-z^2/2} \, dz \right)$$

$$= \frac{1}{\sqrt{2\pi}} \int_{-\infty}^{\infty} e^{-z^2/2} \, dz$$

$$= 1 \text{ from Property 2 (shown below)}$$

To Indicate That the Area Equals 1

We want to show that

$$\int_{-\infty}^{\infty} \frac{1}{\sqrt{2\pi}} e^{-z^2/2} \, dz = 1$$

i.e.

$$\int_{-\infty}^{\infty} e^{-z^2/2} \, dz = \sqrt{2\pi}$$

The mathematics required for this integration is too advanced for A-level, but we can carry out a numerical integration using Simpson's rule. Take ordinates from $z = -4$ to $z = +4$ as the area outside this interval is very small:

$$\frac{1}{3} [(y_{-4} + y_4) + 4(y_{-3} + y_{-1} + y_1 + y_3) + 2(y_{-2} + y_0 + y_2)]$$

$$= \frac{1}{3} [(e^{-8} + e^{-8}) + 4(e^{-4.5} + e^{-0.5} + e^{-0.5} + e^{-4.5}) + 2(e^{-2} + e^{-0} + e^{-2})]$$

$$= \frac{1}{3} [2e^{-8} + 8e^{-4.5} + 8e^{-0.5} + 4e^{-2} + 2e^{-0}]$$

$$= \frac{1}{3} [2(0.000\,335) + 8(0.011\,109) + 8(0.606\,531) + 4(0.135\,335) + 2]$$

$$= \frac{7.483\,129}{3}$$

$$= 2.494\,376$$

As $\sqrt{2\pi} = 2.506\,628$, we have a result which is reasonably close, given the range of values chosen and the method of working.

Alternatively, using the trapezium rule from $z = -6$ to $z = +6$ gives

2.507 404 6 with   8 divisions in the range
2.506 628 26 with 16 divisions in the range
2.506 628 27 with 32 divisions in the range

The General Form of the Normal Distribution

Starting with $Z \sim N(0,1)$, we have the diagram in Figure 5.13.

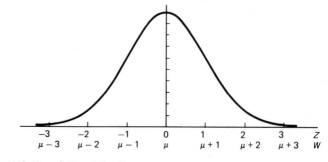

Figure 5.13 $Z \sim N(0,1)$ and $W \sim N(\mu,1)$

A distribution $W \sim N(\mu, 1)$ will be based on the midpoint $x = \mu$; so we have the same curve $(1/\sqrt{2\pi})e^{-z^2/2}$ but with $z = x - \mu$. Therefore, $W \sim N(\mu, 1)$ has the density function

$$f(W) = \frac{1}{\sqrt{2\pi}} e^{-(x-\mu)^2/2}$$

Similarly, starting with $Z \sim N(0,1)$, we have the diagram in Figure 5.14.

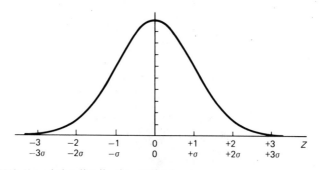

Figure 5.14 $Z \sim N(0,1)$ and the distribution $N(0,\sigma)$

A distribution $N(0,\sigma)$ will be centred at the same place but will have $y = \sigma$, 2σ at $z = 1$, 2 etc.

This would give a total area under the curve of σ instead of 1; so, to make the area equal to 1, we have to scale the vertical axis $f(y)$ by $1/\sigma$:

$$z = \frac{y}{\sigma}$$

So, if $Y \sim N(0,\sigma^2)$,

$$f(y) = \frac{1}{\sigma\sqrt{2\pi}} e^{-(1/2)(y/\sigma)^2}$$

Putting the two ideas together, we get that, if $X \sim N(\mu, \sigma^2)$, then

$$f(x) = \frac{1}{\sigma\sqrt{2\pi}} e^{-(1/2)[(x-\mu)/\sigma]^2}$$

Section **5.6** **Using Tables of the Normal Distribution Function**

As integration of the function

$$f(z) = \frac{1}{\sqrt{2\pi}} e^{-z^2/2}$$

is not possible analytically, numerical tables have been calculated for this standard normal curve, giving the area to the left of any positive values of z, i.e. the (cumulative) distribution function.

This is denoted by the symbol $\Phi(z)$.

Most tables give the area from $-\infty$ to the standardised value of z (for positive values above the mean (Figure 5.15).

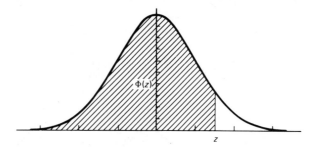

Figure 5.15 $\Phi(z)$ from $-\infty$ to the standardised value of z

Do look carefully, however, at the tables that you are using. Some tables give the area between the mean and z (Figure 5.16); so you will need to add 0.5 to find the value of $\Phi(z)$ for the total area from $-\infty$.

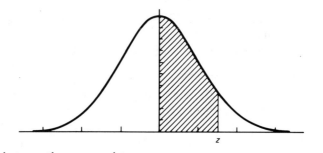

Figure 5.16 $\Phi(z)$ between the mean and z

A few examination board tables give the area to the *right* of positive values of z (Figure 5.17).

This may be more useful for certain types of problem but can cause confusion as the majority of tables are of the first type.

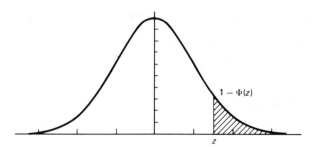

Figure 5.17   Area to the right of positive values of z

We have established that, if $X \sim N(\mu, \sigma^2)$ and $Z = (X - \mu)/\sigma$, then $Z \sim N(0, 1)$. This enables us to use the standard normal tables to find probabilities for any normal distribution, e.g. for $P(X < a) = P(Z < (a - \mu)/\sigma)$; thus the answer can be read from tables of $\Phi(z)$.

Section **5.7   Poisonous or Edible? Can we Solve the Problem?**

The heights of the fungi *Hypholoma capnoides* and Sulphur Tuft can be modelled by normal distributions. Figure 5.18 shows how the two overlap.

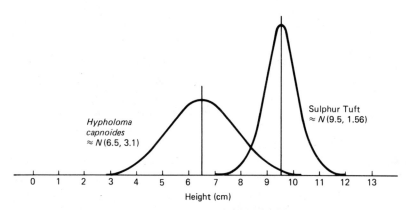

Figure 5.18   Normal distributions for *Hypholoma capnoides* and Sulphur Tuft

Hypholoma capnoides has a mean height of 6.5 cm and a variance 3.1 cm, and Sulphur Tuft has a mean of 9.5 cm and a variance of 1.56 cm. Can we use these population models to help us to distinguish between the two?

Do Many *Hypholoma capnoides* Grow above 8 cm?

Would this be a useful guide to help us reject the poisonous variety?

Using the normal distribution as a model, we can find the probability that a *Hypholoma capnoides* grows taller than 8 cm. To find the proportion of edible fungi taller than 8 cm, we need the area to the right of that value in Figure 5.19.

In order to use tables first convert the variable X to a standardised normal variable Z. We need to know how many standard deviations above the mean is 8 cm.

$$Z = \frac{8 - 6.5}{1.76}$$

$$= 0.852 \qquad \text{(Figure 5.20)}$$

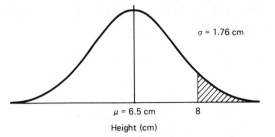

Figure 5.19 Heights of *Hypholoma capnoides* $\sim N(6.5, 3.1)$

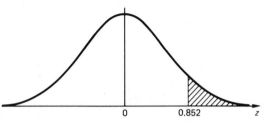

Figure 5.20 $Z \sim N(0, 1)$

We can now find the area to the left of $Z = 0.852$ from tables:

$\Phi(0.852) = 0.8029$

If we subtract this value from 1, we can calculate the area to the right of $Z = 0.852$. This gives a probability of 0.1971. Thus, nearly 20% of edible fungi grow taller than 8 cm. Thus, if we pick fungi shorter than 8 cm, we shall be rejecting about 20% of the population of edible *Hypholoma capnoides*.

As we are more worried about the possibility of picking a poisonous fungus by mistake, we should calculate the probability that a mature Sulphur Tuft is shorter than 8 cm.

Are Many Poisonous Sulphur Tufts Shorter than 8 cm?

The heights of Sulphur Tufts can be modelled by a normal distribution with a mean 9.5 cm and variance 1.56 cm (Figure 5.21).

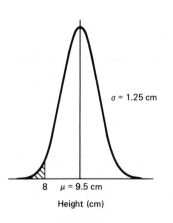

Figure 5.21 Heights of Sulphur Tufts: $\sim N(9.5, 1.56)$

To find the area to the left of 8 cm, we must know how many standard deviations below the mean this is.

$$Z = \frac{8 - 9.5}{1.25}$$

$$= -1.2$$

To use the normal tables for values of $Z < 0$, we must take into account the symmetry of the normal distribution (Figure 5.22).

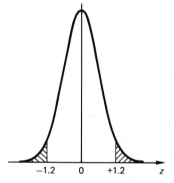

Figure 5.22 $P(Z < -1.2) = P(Z > +1.2)$

The area to the left of $Z = -1.2$ equals the area to the right of $Z = +1.2$:

$\Phi(-1.2) = 1 - \Phi(1.2)$

$\Phi(1.2)$ $= 0.8944$ (area to left of $Z = 1.2$)

$1 - \Phi(1.2)$ $= 0.1056$ (area to right of $Z = 1.2$)

$\Phi(-1.2) = 0.1056$ (area to left of $Z = 1.2$)

So, approximately $10\frac{1}{2}\%$ of Sulphur Tufts will be shorter than 8 cm (quite a large percentage!). In this situation, it will be safer for us to pick fungi which are even shorter than 8 cm even though we shall be rejecting a larger proportion of edible fungi.

Can you solve this problem? What proportion of Sulphur Tufts will be shorter than 7 cm? Is this a 'safe' height to distinguish the edible varieties from the poisonous ones?

References

Klan, J., 1981, *Hamlyn Colour Guide to Mushrooms and Fungi*, Hamlyn, London.
Wallace, D., 1980, Distinguishing Little and Reed Buntings, in J. T. R. Sharrock (ed.), *Frontiers of Bird Identification*, Macmillan, London and Basingstoke.

Chapter 6 Taking Samples

Section 6.1 Finding Answers

By using and studying statistics, we hope to find understandable patterns in complex situations. We hope to find the answers to questions by collecting information ourselves or by analysing data already available. Collecting our own data raises many questions.

For example, can we question people directly for the information? If we do, will our respondents tell the truth, or indeed do they even know the truth?

If we use information collected by another person, we must consider the methods that they have used in order to judge the accuracy and reliability of their results.

Practical 6.1 Heights

1. Ask each person in the class to write down their own height. This may be more difficult than you might expect, particularly if you ask for metric measurement.
2. Check the accuracy of the answers by measuring each person's height. This will give better results but, even so, how accurate can your measurements be? Is it possible or even desirable to measure heights to the nearest millimetre? Would measurements to the nearest centimetre be better?
3. Discuss your findings. Did people tend to underestimate or overestimate? Was this tendency different among boys and girls?

You now have the distribution of heights for the students in your class. This task was fairly straightforward to accomplish, but how would you cope with finding the distribution of heights for all 16–17 year olds in Britain, or even in your town? Clearly, some kind of sample would need to be taken to make the task manageable.

Section 6.2 Why Take a Sample?

The statistician's work generally begins with a problem, e.g. 'Are today's teenagers taller than their parents?' or 'Do teenagers today have more spending money than their parents did?'. Both problems ask a question about two groups of people: teenagers and parents. Each of these is called a target population and can be identified by some common characteristic, in this case age. However, defining these two populations precisely is not as easy as it might first appear. Firstly, who are teenagers? Do we mean all people aged 13–19 years?

If we are considering the first question, do we want to include younger teenagers who may still be a long way from their adult height? Also, as the growth rates of boys and girls are so different, they may need to be considered separately. For 'parents', are we going to compare our group of teenagers with all adults, with adults having teenage children, or with their own parents?

Having defined the population(s) of concern, we then have to decide on the population characteristic or variable to be measured, and the method of obtaining the data. For example, should we measure the heights of our adult population now, or should we try to find information on their heights when they were teenagers? This latter alternative seems fairer, but the information will be difficult to obtain.

If we consider the question about spending money, we find even more problems with measurement. In defining spending money, should we include income from part-time jobs or pocket money only? Should we include teenagers who have left school, and should we adjust their income by subtracting payments for board and lodging? Will parents remember the amounts of pocket money that they received, or how much they contributed to family housekeeping? There is also the problem of how we take account of inflation.

For each of the problems under consideration, it clearly is not possible to obtain measurements from every member of both our two target populations. Because of lack of time and resources, we have to compromise by selecting from each population group a smaller group or sample. It is in our interests to do everything possible to ensure that our sample is representative of the population, by trying to eliminate any source of bias in their selection. We must not exclude sections of the population from our sample. For instance, we must not restrict our sample to particular income groups, or to people living in urban areas, as clearly either of these factors could be an important influence on the results of either problem. By giving every member of the population an equal chance of being included in the sample, we can eliminate any deliberate source of bias. Samples conducted in this way are called random samples.

Lotteries are sometimes used to select a group of people in a way which is fair or random. For example, in the USA, lotteries have been used to determine the draft for call-up into military service in times of war. The method used involved selecting dates from the 365 possible dates in a year. Men whose birthdays were on those dates would be called up in order of selection. One particular lottery conducted for the 1970 draft was criticised for being unfair. It later transpired that the lottery capsules had been poured into the box month by month, starting with January. By examining the order in which the capsules were subsequently drawn, it was proved that they had not been mixed adequately. The capsules for December and November had tended to remain towards the top of the box and had been drawn out earlier. So even a lottery has to be very carefully administered to ensure that it is fair.

In the US draft lottery, men were selected and identified according to their birthday. If we wish to use a lottery to select a sample, we are more likely to give each member of the population a unique number of identification. However, this pre-supposes that each member can be identified and listed. Such a list is called a sampling frame, as the sample will be taken from it. Obtaining a complete and accurate sampling frame for the examples that we have quoted is virtually impossible. One of the better ones available for adults is the Electoral Register. This lists households by addresses and gives the names of adults over 18 years in each household. It also includes the name and date of birth of people over $16\frac{1}{2}$ years who will be eligible to vote in that coming year. Unfortunately, as some people change address and as others die, the register is never fully accurate. For some surveys the electoral register would not make a suitable sampling frame.

Discussion Points

Where would you find, or how might you obtain information to draw up, a list of teenagers which would make a suitable sampling frame for the examples given at the beginning of Section 6.2? How might you obtain a list of elderly people?

Section 6.3   **Selecting the Sample**

Once a sampling frame has been found, each member on the list can be identified by a number. A random sample may be chosen by a lottery, as described earlier, or by generating random numbers. Microcomputers and many calculators are able to generate numbers which can be taken as random. Alternatively, tables of random numbers can be used. A table of random numbers is a list containing the ten figures 0–9 in random order so that they have the following properties.

1. The figures are independent in any position in the sense that the value of a figure in any one position has no influence on the value of any other.
2. The figure in any position on the list has the same chance of being any one of the numbers 0–9.

From these first two properties, it follows that, if figures are read in groups of two, each pair has an equal chance of being any one of the 100 possible pairs 00, 01, 02 up to 99 (or, for groups of three, any one of the 1000 possible three-figure numbers from 000, 001 up to 999).

Practical 6.2   **Random Numbers**

Write down 100 random numbers of your own using the numbers 0–9. Do not use a calculator or tables; just see whether you can write down the numbers yourself fairly quickly.

Table 6.1 shows a set of supposedly random numbers and Table 6.2 the frequency distribution of these numbers.

Table 6.1   A set of supposedly random numbers

6	4	2	0	9	8	1	3	5	2
7	3	4	2	5	9	0	2	6	5
8	9	6	7	4	0	3	2	1	5
5	6	3	0	2	1	9	5	3	0
4	2	0	9	8	6	3	7	0	1
5	0	3	4	9	8	3	6	2	4
8	6	2	1	0	3	5	8	6	3
9	0	5	6	3	5	0	1	3	2
2	5	8	6	9	3	2	0	6	5
5	3	0	2	6	1	9	8	6	3

Table 6.2   Frequency distribution of single figures

Figure	Frequency
0	13
1	7
2	13
3	15
4	6
5	13
6	13
7	3
8	8
9	9
Total	100

This set has a poor distribution of single-figure numbers with only three 7s appearing. When considered in pairs, however, certain two-figure numbers seem prone to repetition (i.e. 25 occurs twice, 58 occurs three times and so on). However, our analysis at the end of the chapter shows that the number of repetitions is similar to the number that we might expect.

However, in looking at two-figure numbers, we might consider the number of 'doubles' occurring, 00, 11, 22, etc. In a table of 99 two-figure numbers, we should expect to find 9 or 10 and in fact only 3 occur. These have all occurred as the last number of one line and the first number of the next. The person writing these numbers seems biased against using the same figure in successive places.

Section 6.4 How to Use Random Numbers

Random numbers may be used to select a random sample from a sampling frame. Each member of the population on the list is allocated one identifying number on the list before the selection process begins. Random numbers are then allocated according to the size of the population. For instance, random numbers could be used instead of a lottery for the US draft. As there are 365 days in a year, random numbers would have to be used in groups of three.

The numbers could be allocated thus:

1 January	001
2 January	002
3 January	003
⋮	⋮
30 December	364
31 December	365

Only 365 numbers have been used so far and the numbers 000 and 366–999 would be ignored or rejected if they occurred in the selection process. To eliminate this wastage, each day could be allocated two numbers. However, each item in the sampling frame must have an equal chance of being selected and so we do not have enough random numbers to allocate any more than two to each day.

An uncomplicated scheme such as the following should be adopted:

1 January	001 and 501 (say)
2 January	002 and 502
3 January	003 and 503
⋮	⋮
30 December	364 and 864
31 December	365 and 865

Now, only 270 numbers are left unused, but these must not be allocated as each day must have an equal chance of being selected.

Having allocated numbers to the members of the population, the sample can be selected using a table of random numbers. For example, if the numbers 032, 519, 360 and 948 appeared in the table the days selected would be

033	2 February
519	19 January
360	26 December
948	discard this number

If such a scheme were used to select birth dates for including men in a call-up draft, it would give each day in the year an equal probability of being included. However, it still would not be fair to individuals, as it does not allocate equal probabilities to individuals, only to their dates of birth. Births do not follow a uniform distribution throughout the year, as some months are less popular. Also medical intervention in childbirth means that fewer births are now taking place at weekends and bank holidays, and babies born on Christmas Day are becoming quite rare.

Simple Random Samples

When selecting simple random samples, it is possible either to allow for the same item to be chosen more than once or to make sure that no item does occur more than once. The first method is sometimes called sampling with replacement; at each stage of the sampling, every item has an equal chance of being chosen. This is called simple random sampling with replacement. The second method is sampling without replacement; at each stage of the sampling, every remaining item has an equal chance of being chosen. This is called simple random sampling.

When choosing a sample of size r from a population of size n, then simple random sampling with replacement ensures that each of the possible n^r samples has the same probability of being chosen. Individual items may occur more than once. Simple random sampling ensures that each of the possible ^nC_r samples has the same probability of being chosen. No individual item can occur more than once.

In the remainder of this chapter the term *random sample* will be taken to mean a sample obtained by simple random sampling with or without replacement. If the distinction is important, then the phrase *sampling with replacement* will be used.

Practical 6.3   Comparison of Random and Non-random Samples

Page PC9 contains a population of 60 circles. Select a sample of 5 circles which you think would be representative. Measure the diameter of each circle in your sample. (All diameters are multiples of 0.5 cm.) Calculate the mean of the diameters in your sample. Repeat the procedure for two more samples. Put together the results for the whole class by making a frequency table for your sample means and calculate the average of all the sample means.

When this has been completed, label the circles 01, 02, ..., 59, 60 in any order. Use two-figure random numbers or a calculator or microcomputer to select a sample of 5 circles from the population. Again calculate the sample mean and take two further samples using different parts of the random number table. Compare the class results for the random samples with those for the non-random samples.

Are the means for the random samples generally larger or smaller than those from the non-random samples?

The mean for the population of 60 circles is given at the end of this chapter section. Both sets of samples should be examined to see how close their results were to the true population value.

Which type of sample gave you the best results?

Further Practicals

You could follow through Practical 6.3 using a different population to sample from. Here are two other suggestions.

Practical 6.4   Pebbles

Use a pile of about 40 or 50 pebbles, from which samples of 5 are to be taken and weighed. To take random samples, number each pebble using an adhesive label, find the weight of each pebble in the pile and calculate the population mean. It has been suggested that the largest stone will rarely be included in non-random samples so that the samples will usually underestimate the true mean.

Practical 6.5   **Heights**

Another alternative is to take samples of 5 students in your class and to measure their heights. Follow through the same procedures as outlined for numbers 1 and 2.

Section 6.5   **Other Populations and Samples**

So far, we have considered mainly human populations, and samples from them. However, other types of population may well be of interest to the statistician. For the statistician in industry, the population under consideration may be the output from a production line. It is not possible to test every item because of time and expense. In many cases such as in tasting food or drink, testing fireworks or fuses or finding the breaking strength of steel, testing the item will necessarily destroy it. The quality controller may not use a strictly random sample but may systematically take samples at, say, hourly intervals. A sample could consist, for example, of 5 litres of water taken each hour from a purification plant, a 10 metre length of cloth from each loom inspected at the beginning and end of each day, 1 litre of gas taken at random intervals, and so on.

Other populations may be theoretically infinite. For example, any number of slightly different results are possible if physical measurements of the speed of light are taken. In the classroom, where a die is being thrown, our results may be considered as a sample from an infinite number of possible throws.

Section 6.6   **Why Take Random Samples?**

Each random sample that we take from any population will give a different value for $\bar{x}$, the sample mean. Although this appears very confusing at first, the variability of random samples can be calculated mathematically from the laws of probability. The statistician can predict the sort of variability that is expected from a random sample, but it is often not possible to do this for non-random samples.

In any situation where a sample is being taken, the following ideas need to be kept clear (Figure 6.1). In the real world, we have the population from which the actual sample is

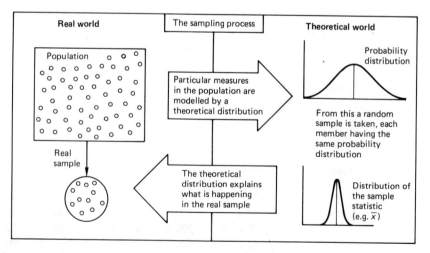

Figure 6.1   Comparison of the real world with the theoretical world

drawn. In parallel to this, the measurement of interest in the population (height, age, income, etc.) will be modelled by a theoretical probability distribution. The equivalent to sampling in the real world from the real population is the drawing of samples from the distribution in the model. Random sampling enables us to calculate sample variability when samples are taken from the theoretical distribution. Therefore the type of sampling that we use in the real world needs to match up with this so that reliable probabilities and estimates can be calculated.

While the variability of random samples is calculable and predictable, the statistician must be careful that the practical design of his or her study does not have hidden factors which destroy the randomness of his sample. The predictable variability which can be found among simple random samples is called sampling error. For example, we know that every now and again we can expect to get a sample of all low values, which will seriously underestimate our population mean just 'by the luck of the draw'. However, we can predict just how often this will occur. It is much harder to cope with the problems of bias caused by improper sampling. These are called non-sampling errors. They encompass a variety of sources of bias. If an incorrect sampling frame is used, the sample selected from it will not be a random sample from the target population. If some members of the population are not listed, they have a zero probability of being included in the sample.

Bias may also arise later as the study progresses and surveys of human populations are most vulnerable. If a questionnaire study is used, some of the sample members may forget or refuse to fill it in. The 'non-respondents' may in fact be an important group with different characteristics from the rest of the population. Similarly, bias is introduced if volunteers are used for a sample.

If a survey makes use of an interviewer, their race or sex has been found to influence the results in many cases, even if standard questions are used. Even if questions are unambiguous, and avoid leading the respondent to particular answers, he or she may make errors in answering them. The respondent may even falsify answers to sensitive questions (even quite simple ones such as age). Finally, yet more errors may be introduced when the responses are coded for analysis.

Section 6.7 Stratified Sampling

Practical 6.6 Favourite Television Programmes

In situations where there may be diverse groups within a population, stratified sampling may be used. This is a modification of random sampling. The statistician must first decide on what groupings do exist within the population, and the proportion of the total population in each group. Separate random samples are then taken from each of the strata (or groups). In stratified sampling, it is usual, but not absolutely necessary, to have the strata from the population represented proportionately in the sample. The strata must be completely distinct and must cover all the population. Here is an example showing how stratified sampling may be carried out.

In a mixed upper school there are a total of 1000 pupils in the third, fourth and fifth years. The pupils are asked to state their favourite television programme and their reply is to be categorised as one of the following.

A Children's programmes.
B Pop programmes.
C Documentaries, current affairs and news.
D Sport.
E Serials and soap operas.

Table 6.3   Favourite television programmes of a mixed upper school

Third year; 400 pupils					
DACDEAAADB	AAADABBADE	BDCBACADEA	BAECAABABD	DEBABABAAA	1–50
ABCBDAEBBB	ACBAAABBED	BCDAAADCAA	BAECADEBAD	CBABABBABB	51–100
AADBBDAABE	EABCBBBBAA	ADCBABDBEB	AADDDACDBC	CBEBCEDCBE	101–150
BCBBAABCED	BBBABDEDDE	BBBBABECCA	BCBDBBBEBB	BBACDDABAE	151–200
BABBDAEADB	BBABDAACBE	AADEABEBCD	AABDEADDAB	CBBADABAAB	201–250
AADBADDBDB	BBAABBAABA	DBABAAAADE	DDDBAADCBB	BADCACEDBA	251–300
BABBDCAACD	AADDBDBBAB	BCEDBDABED	ACADAEBADA	BECDDACDCB	301–350
CDBAADEAAD	DCDDEBDABA	DDABDBDCEE	ADAAEBEEAA	EDAADBCEED	351–400
Fourth year; 300 pupils					
CDCEBDEBDD	CBDECBDBCB	CDDECCBCAD	ADEEDABBDD	BBBEBDDBAB	401–450
CDDDBEDBDE	BCDEEBBDBB	BBADBDBACB	BCEDBEDEDD	BDEECCDEBC	451–500
EAEDBECDCD	BDDDBEBDDD	CDBDEADEAB	EEDBBCDEBD	EEDDBDBEEA	501–550
EBCBDBBCBB	AEEBDCCDDD	DCDDBEEBEC	ECEBEBBCCD	DBEDDCDDDB	551–600
BEADEDEBBB	EDDBDEDCEB	BBDCDBBDBE	BBECBECEBD	CCDECDCBDB	601–650
CDDBCBABCD	ECCEBBBCBB	DDABCDEBDE	DEBDEEBBAB	DEDDBDCBEE	651–700
Fifth year; 300 pupils					
DEBCCEBBDD	EDBBDEEEBD	DEBCECDBBD	EDEDBEEEEE	EBEBDBEECB	701–750
BDDBDDBBEB	EBDDBEBEBD	EBEDEBBDED	EDEBDBDDBD	BEBEDDBBDE	751–800
EBDBEBDBEE	DDBDDBBBBD	EBBDBDEDED	EDEDEDBEEC	BDDEBEECEB	801–850
EEBBEEEDBE	EDEBDBBDDC	BBBCDDBCDD	BDEEDDBBDE	CBDDBDBBBD	851–900
DEDBEBDBEE	BEEDDBEBEB	EDDBEEBBDE	EDDEDDBEEB	EDDBBBDBBB	901–950
EDEBBEBBBC	BEBEDEEBEB	BBEEDDCCBD	BDBBEBDEBB	BBCBDBEDEE	951–1000

The replies for the whole school are listed in Table 6.3.

1. Firstly, use random numbers to choose a random sample of 100 pupils from the entire school. Using tally marks, record the number of pupils choosing programmes in each category.
2. Next select a stratified sample of 100 pupils. The population is to be stratified by year groups as it is reasonable to expect that pupils of different ages may select different types of programme. Each year group will then be represented proportionally in the sample according to its size. In the third year there are 400 pupils out of the total of 1000. The stratified sample will contain (400/1000) × 100 third-year pupils, i.e. 40 pupils.
3. Calculate the numbers of fourth- and fifth-year pupils in the sample, and then select your sample using random numbers.
4. Find and record the total number of votes for each category from the sample using Table 6.4 on page 129.
5. Find the actual number of votes for each group by counting all the results (share this task among the class).
6. Of your 2 samples, which type of sample was the more representative?

Section **6.8**   **Sampling from Finite Populations**

If a small population is considered, it is possible to enumerate all the possible samples which could be taken. By doing this, we can find all the possible values for any sample

Table 6.4   Total number of votes for each category

Year	Number of choices in this category				
	A	B	C	D	E
Third Fourth Fifth					
Total					

Table 6.5   Frequency of coins

Coin	Frequency
1 p	3
2 p	3
5 p	4

statistic that we choose, e.g. the mean, the median or the variance. In this section, we shall concentrate on the means of samples of size 2 taken at random from 10 coins whose values are given in Table 6.5.

The mean and the variance of this small population are 2.9 p and 3.09 p, respectively.

Sampling with Replacement

If samples of size 2 are taken with replacement, every single possible value of the mean can be calculated with the corresponding probabilities. As the first coin is to be replaced, the probabilities at the second selection remain the same (Table 6.6).

Table 6.6   Probability distribution of $\bar{X}$ for samples of size 2

$\bar{x}$	Associated sample values		Probability	
1	1 and 1		0.3×0.3	$= 0.09$
$1\frac{1}{2}$	1 and 2	2 and 1	$0.3 \times 0.3 + 0.3 \times 0.3$	$= 0.18$
2	2 and 2		0.3×0.3	$= 0.09$
3	1 and 5	5 and 1	$0.3 \times 0.4 + 0.4 \times 0.3$	$= 0.24$
$3\frac{1}{2}$	2 and 5	5 and 2	$0.3 \times 0.4 + 0.4 \times 0.3$	$= 0.24$
5	5 and 5		0.4×0.4	$= 0.16$
		Total		$= 1.00$

From this probability distribution for $\bar{X}$, the mean and variance for $\bar{X}$ can be calculated (Table 6.7).

Table 6.7   Calculation using $p(\bar{x})$

$\bar{x}$	Probability $p(\bar{x})$	$\bar{x}p(\bar{x})$	$(\bar{x})^2 p(\bar{x})$
1	0.09	0.09	0.09
1.5	0.18	0.27	0.405
2	0.09	0.18	0.36
3	0.24	0.72	2.16
3.5	0.24	0.84	2.94
5	0.16	0.80	4.00
Total		2.90	9.955

Therefore,

$$\text{mean of } \bar{X} = \sum \bar{x} p(\bar{x}) = 2.9$$

$$\text{variance of } \bar{X} = \sum (\bar{x})^2 p(\bar{x}) - \left(\sum \bar{x} p(\bar{x}) \right)^2$$

$$= 9.955 - (2.9)^2$$

$$= 9.955 - 8.41$$

$$= 1.545$$

Note that the mean of $\bar{X}$ for samples of size 2 is the same as the population mean, but the variance is one half.

Exercise 6.1   **Values of $\bar{X}$ and their Probabilities**

Using Table 6.8 to help you, find all the possible values of $\bar{x}$ and their probabilities, if samples of size 3 are taken from the same population.

Table 6.8   Probability distribution of $\bar{x}$ for samples of size 3

$\bar{x}$	Associated samples	Probability
1	(1, 1, 1)	$0.3 \times 0.3 \times 0.3 \quad = 0.027$
$1\frac{1}{3}$	(1, 1, 2)(1, 2, 1)(2, 1, 1)	$(0.3 \times 0.3 \times 0.3) \times 3 = 0.081$
$1\frac{2}{3}$		
2		
$2\frac{1}{3}$		
$2\frac{2}{3}$		
3		
$3\frac{2}{3}$		
4		
5		
Total		≈ 1

Using the probability distribution for $\bar{X}$, calculate the mean and variance for $\bar{X}$. What do you notice about these values? For simple random sampling with replacement from a distribution with mean μ and variance σ^2, theory predicts that the mean of $\bar{X}$ should be μ and the variance of $\bar{X}$ should be σ^2/n, where n is the sample size. Do your results confirm this?

Sampling Without Replacement (Simple Random Sample)

Suppose that the samples are taken without replacement, i.e. the first coin is not replaced before the second is selected. Now the probabilities for the second selection will change and are conditional upon the result of the first selection (Table 6.9); when considering small samples taken from a small population, it is possible to calculate all the conditional probabilities.

From the probability distribution of $\bar{X}$, the mean and variance of all the possible values of $\bar{X}$ can be calculated (Table 6.10).

Table 6.9   Probability distribution of $\bar{X}$ for samples of size 2

$\bar{x}$	Associated sample values	Probability
1	(1, 1)	$\dfrac{3}{10} \times \dfrac{2}{9} = \dfrac{6}{90}$
$1\frac{1}{2}$	(1, 2)(2, 1)	$\dfrac{3}{10} \times \dfrac{2}{9} \times 2 = \dfrac{18}{90}$
2	(2, 2)	$\dfrac{3}{10} \times \dfrac{2}{9} = \dfrac{6}{90}$
3	(1, 5)(5, 1)	$\dfrac{3}{10} \times \dfrac{4}{9} \times 2 = \dfrac{24}{90}$
$3\frac{1}{2}$	(2, 5)(5, 2)	$\dfrac{3}{10} \times \dfrac{4}{9} \times 2 = \dfrac{24}{90}$
5	(5, 5)	$\dfrac{5}{10} \times \dfrac{3}{9} = \dfrac{12}{90}$
	Total	$\dfrac{90}{90} = 1$

Table 6.10   Calculation using $p(\bar{x})$

$\bar{x}$	$p(\bar{x})$	$\bar{x}p(\bar{x})$	$(\bar{x})^2 p(\bar{x})$
1	$\dfrac{6}{90}$	$\dfrac{6}{90}$	$\dfrac{6}{90}$
$1\frac{1}{2}$	$\dfrac{18}{90}$	$\dfrac{27}{90}$	$\dfrac{81}{180}$
2	$\dfrac{6}{90}$	$\dfrac{12}{90}$	$\dfrac{24}{90}$
3	$\dfrac{24}{90}$	$\dfrac{72}{90}$	$\dfrac{216}{90}$
$3\frac{1}{2}$	$\dfrac{24}{90}$	$\dfrac{84}{90}$	$\dfrac{204}{90}$
5	$\dfrac{12}{90}$	$\dfrac{60}{90}$	$\dfrac{300}{90}$
	Total		$\dfrac{1761}{180}$

Therefore,

$$\text{mean of } \bar{X} = \sum \bar{x}p(\bar{x})$$

$$= \frac{261}{90}$$

$$= 2.9$$

$$\text{variance of } \bar{X} = \sum (\bar{x})^2 p(\bar{x}) - \left(\sum (\bar{x}) p(\bar{x}) \right)^2$$

$$= \frac{1761}{180} - (2.9)^2$$

$$= 9.783 - 8.41$$

$$= 1.373$$

Exercise 6.2   **Sampling without Replacement**

If sampling is done without replacement, calculate the probabilities for all the possible values of $\bar{X}$ if samples of size 3 are taken. Then find the mean and variance of that probability distribution.

For simple random sampling from a population of size N with mean μ and variance σ^2, theory states that the mean of $\bar{X}$ should be μ and the variance of $\bar{X}$ should be

$$\frac{\sigma^2}{n} \frac{N-n}{N-1}$$

where n is the sample size. Do your results confirm this?

Repetitions of Two-figure Numbers (i.e. 00–99) in Practical 6.3

Here are the two-figure numbers which were repeated in the 100 numbers written down by a person completing Practical 6.2:

2 repetitions	01, 09, 13, 15, 19, 25, 26, 34, 50, 53, 62, 65, 90.
3 repetitions	02, 03, 20, 21, 30, 32, 35, 42, 55, 58.
4 repetitions	98.
5 repetitions	63, 86.

While this might appear worrying at first, the distribution of repetitions can be analysed as following an approximately Poisson distribution.

There are 99 possible positions on the list in which a two-figure number might appear. There are 100 possible two-figure numbers so that $n = 99$ and $p = 1/100$. If we use the Poisson distribution as a model (Table 6.11), the mean will equal $99/100$.

Table 6.11   Poisson probabilities

Number of appearances	Poisson probabilities	Expected frequencies
0	0.3716	36.79
1	0.3679	36.42
2	0.1821	18.03
3	0.0601	5.95
4	0.0149	1.48
5	0.0029	0.29

This shows that the number of appearances is not too different from what we might expect.

Solution to Practical 6.3

Circles: mean diameter = 1 cm.

Chapter 7 Estimating Proportions

Section 7.1 Can Election Results be Predicted?

The *News Chronicle* commissioned Gallup to conduct Britain's first constituency poll in 1938 during the West Fulham by-election. This poll predicted that Labour would win and in fact was less than 1% out in the final result. 46 years later at the Fulham 1986 by-election the number of polls had proliferated and all the major polling organisations were close in their prediction of the result (Figure 7.1).

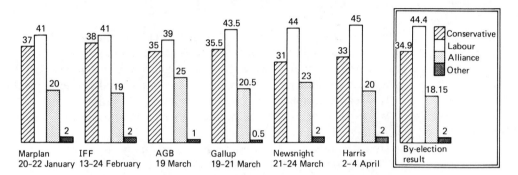

Figure 7.1 The opinion polls and the final result in the Fulham by-election on 10 April 1986

Occasionally, mistakes are made; a Labour victory predicted at the Brecon and Radnor by-election in 1985 finally materialised as a Liberal win! However, pollsters try to avoid such disasters if at all possible. Not only is a major mistake embarrassing for the company concerned, but also it could have repercussions for their commercial contracts in market research.

Investigate a recent election or by-election, and find the details of as many political opinion polls as you can. Compare their estimates of the proportion of the electorate voting for each of the main political parties. If the election result is known, see how close each of the poll predictions were.

Section 7.2 Conducting an Opinion Poll

Practical 7.1 Opinion Poll

Conduct an opinion poll using a random sample, or stratified random sample of 100 pupils in your school. (The task can be shared among the students in your group.) Choose a question on a topic which interests you and which can be answered as a straight yes or no reply. The question must not embarrass your respondents as you can maximise the accuracy of their replies by ensuring that there is no reason for them to hide the truth. Here are some examples which you might use or adapt.

1. Are you a vegetarian?
2. Do you think that the wearing of seat belts should be compulsory for passengers in the back seat of cars?

3. Should 16 year olds be allowed to drink alcohol in pubs?
4. Should the ice-cream van at the school gate be allowed to continue?
5. Do you have a driving licence?

Try to make sure that the question you use will be answerable by the pupils or students in your survey. It should be on a subject about which they have direct knowledge or about which they can be expected to have an opinion.

Choose your sample of 100 and conduct your survey. Record your results simply by totals of yes and no replies.

Section 7.3   The Accuracy of the Results

How do opinion polls arrive at the right answer? *The Guardian* Marplan Index is a national poll conducted monthly using a 'representative quota sample' of about 1400 adults aged 18 + years in over 100 randomly selected constituencies. On average, this works out at 14 adults per constituency—a tiny proportion of the total voting population.

How can a sample of this size be adequate for a national poll?

What sized sample will be needed in a single constituency by-election poll to get the same level of accuracy?

How accurate will the estimates from our survey be, based on a sample of 100?

These questions will be considered in this section.

The Observer Harris prediction that 20% of the Fulham voters would vote for the Alliance was borne out in the final result. Yet, at the time of writing, predicting the proportion of Alliance voters would seem to be more problematical than those for Conservative or Labour, particularly in constituencies where the Alliance is the third party. The problem appears to be that it is difficult to distinguish between voters who positively support a given party and those whose main interest is to defeat one of the parties. This second type of voter will vote for whichever of the two remaining parties seems most likely to defeat the disliked party. To a certain extent, the opinion polls themselves exacerbate this process and will frequently encourage 'floating' voters to change allegiance at the last minute.

If the predictions for Conservative and Labour votes are classified together as 'non-Alliance votes' the problem can be viewed as a binomial situation. The opinion pollster has to estimate p, the proportion of Alliance voters in the population. By interviewing a sample of n voters the number A of Alliance voters in the sample can be used to gain an estimate of p.

Thus, p is the unknown proportion of Alliance voters in the population, and also p is the probability that a voter picked at random will be an Alliance voter; n is the sample size and A is the number of professed Alliance voters in the sample.

A can be viewed as being distributed binomially: $A \sim B(n, p)$. The opinion pollster aims to estimate p when A is known for a given sample but also wants to have some measure of the accuracy of this estimate.

For example, the pollster may wish that his or her estimate of the proportion be within 0.01 or 1% (represented diagrammatically in Figure 7.2).

More generally, the value of 1% can be replaced by e to stand for an error in the proportion of any given magnitude which can be decided in advance (Figure 7.3).

The obvious estimate to use is A/n, the sample proportion, but it is not possible to be absolutely certain that the true value of p lies within a given error of this sample value.

If we cannot be absolutely certain that the true value of p lies within a given range of

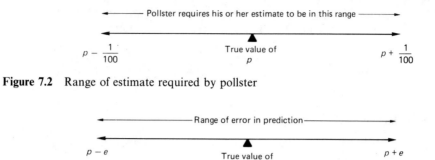

Figure 7.2 Range of estimate required by pollster

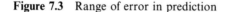

Figure 7.3 Range of error in prediction

this sample estimate, then the next best might be to have this happen with a given probability (e.g. 0.95, 0.99, etc.). This again turns out to be impossible without making many more assumptions about p, but we can do something similar. Instead of a probability interval, we can derive a confidence interval. It is possible to calculate a range such that for a known value of p the sample estimate lies within a given range of p with a given probability (Figure 7.4).

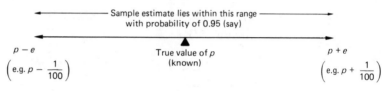

Figure 7.4 Solvable situation

In any real situation, the true value of p will not be known. The pollster would like to be able to construct an interval around his or her sample value A/n, which he or she hopes will include the true value of p with a probability of 0.95 (say).

The estimate to be used will be A/n and we need to know the probability that our estimate occurs in the range $p \pm 0.01$ (or $p \pm e$), i.e. we require the probability that (Figure 7.5)

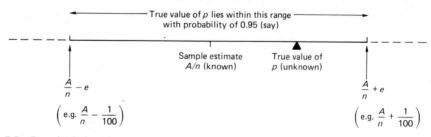

Figure 7.5 Required situation

$$p - \frac{1}{100} < \frac{A}{n} < p + \frac{1}{100}$$

or

$$p - e < \frac{A}{n} < p + e$$

If we multiply throughout by n, we get an expression for the number of successes in our sample, i.e. the number of professed Alliance voters:

$$np - \frac{n}{100} < A < np + \frac{n}{100}$$

or

$$np - ne < A < np + ne$$

We now require the probability that the number of voters in the sample should lie in a certain range (Figure 7.6).

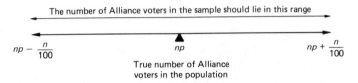

Figure 7.6   The probability that the number of Alliance voters in the sample should give a proportion within ± 0.01 of the true value

More generally, we require that the number of successes in the sample should lie in a certain range (Figure 7.7).

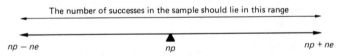

Figure 7.7   The probability that the number of successes in the sample should give a proportion within $\pm e$ of the true value

If n, our sample size, is sufficiently large, the binomial distribution used to model the number A of Alliance voters (or successes) in the sample may be approximated by the normal distribution. Thus, we believe that the number of Alliance voters occurring in a large number of poll samples should follow a normal distribution, with most polls giving answers close to the expected population value. A few polls will give extreme results at each end of the distribution. Our problem now is to find the probability shaded in Figures 7.8 and 7.9.

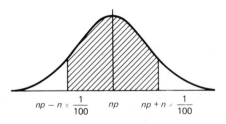

Figure 7.8   Probability that proportion is within ± 0.01 of the true value: normal distribution $N(np, npq)$

The normal distributions in Figures 7.8 and 7.9 can be standardised by dividing the distance away from the mean by the standard deviation which is $\sqrt{npq}$ (Figures 7.10 and 7.11). So $(A - np)/\sqrt{npq} \sim N(0, 1)$.

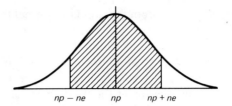

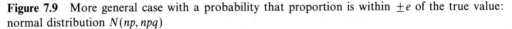

Figure 7.9   More general case with a probability that proportion is within $\pm e$ of the true value: normal distribution $N(np, npq)$

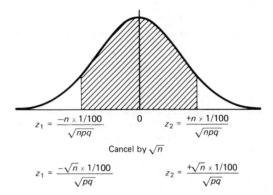

Figure 7.10   Opinion poll with a probability that proportion is within ± 0.01 of true value: $N(0, 1)$ distribution

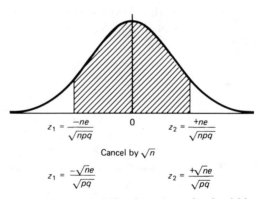

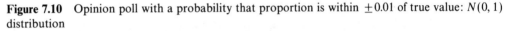

Figure 7.11   General example with a probability that proportion is within $\pm e$ of true value: $N(0, 1)$ distribution

In determining the probability that A lies within the required range, we shall need to calculate the area under the standard normal curve between z_1 and z_2. The researcher may determine the values of n and e, but the values of z_1 and z_2 are determined by the true value of p.

Suppose the true value of p is 0.3 and we take a sample of size 1000 and we want our proportion to be within 0.01 or $1/100$, then we have

$$z_1 = -\frac{\sqrt{1000} \times 1/100}{\sqrt{0.3 \times 0.7}} = -0.69$$

$$z_2 = +0.69$$

The probability (from the standard normal tables) that the sample proportion lies within 0.1 of the true value of p is 0.51, i.e. 51% of all samples drawn from this population will have sample proportions within 0.01 of the true population proportion of 0.3.

An alternative way of approaching this problem is first to fix the probability of being within range and then to calculate the range associated with that probability. This is done later in this chapter under the heading of confidence intervals.

Now, in fact we do not usually know the true value of p, but we can still do similar calculations. By examining what happens for different values of p we can calculate the maximum value of $\sqrt{pq}$.

Using this maximum value of $\sqrt{pq}$, we get an underestimate of the true probability that the sample proportion is in the given range (Table 7.1).

Table 7.1 Probability that the sample proportion is in the given range

p	0.05	0.1	0.2	0.3	0.4	0.5	0.6	0.7	0.8	0.9	0.95
q	0.95	0.9	0.8	0.7	0.6	0.5	0.4	0.3	0.2	0.1	0.05
pq	0.0475	0.09	0.16	0.21	0.24	0.25	0.24	0.21	0.16	0.09	0.0475
$\sqrt{pq}$	0.22	0.30	0.40	0.46	0.49	0.50	0.49	0.46	0.40	0.30	0.22

From this table it can be seen that the maximum value for $\sqrt{pq}$ is 0.5 which occurs when p and q are both 0.5 (and that it is within 10% of this for the range $0.3 < p < 0.7$).

Substituting this in to our expressions for z_1 and z_2, we get, when there is 1% error each side (Figure 7.12),

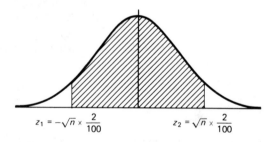

$$z_1 = -\sqrt{n} \times \frac{2}{100} \qquad z_2 = \sqrt{n} \times \frac{2}{100}$$

Figure 7.12 $N(0, 1)$ with $p = 0.5$ showing error in proportion of 0.01 each side

$$z_1 = -\frac{\sqrt{n} \times 1/100}{\sqrt{pq}}$$

$$z_1 = -\frac{\sqrt{n} \times 1/100}{1/2}$$

$$z_1 = -\sqrt{n} \times \frac{2}{100}$$

and

$$z_2 = \sqrt{n} \times \frac{2}{100}$$

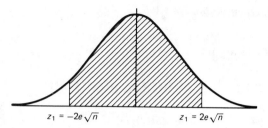

Figure 7.13 $N(0, 1)$ with $p = 0.5$ showing error in proportion of e each side

and, when there is any value of error in the proportion of e each side (Figure 7.13),

$$z_1 = -\frac{e\sqrt{n}}{\sqrt{pq}}$$

$$z_1 = -\frac{e\sqrt{n}}{1/2}$$

$$z_1 = -2e\sqrt{n}$$

and

$$z_2 = 2e\sqrt{n}$$

Thus the minimum probability that the error of the estimate lies within a particular range is the area under the normal curve between z_1 and z_2.

(Again an alternative is to fix the probability and then to obtain an overestimate for the range; this we do in a later section.)

Worked Example

If an opinion poll canvasses 1000 people, what is the minimum probability that the proportion of voters for a particular party is accurate to within $1/100$?

Solution

$$n = 1000$$

$$\sqrt{n} = 31.62$$

$$z_1 = -\sqrt{n} \times \frac{2}{100}$$

$$= -0.632$$

$$z_2 = \sqrt{n} \times \frac{2}{100}$$

$$= 0.632$$

From normal tables the area to the left of z_2 is 0.7373 and the area to the left of z_1 is 0.2627. Therefore, the area between z_1 and z_2 is 0.4746. So at least 47.5% of all polls of size 1000 will have a sample proportion within 1% of the true proportion. (Compare this with the correct value of 51% calculated previously for a known value of $p = 0.3$.)

Practical Follow-up **7.1**

You have conducted an opinion poll using a sample size of 100. Calculate the minimum probability that the proportion of 'yes' voters in the sample is accurate to within 1%.

Section **7.4**   **Simulation of an Opinion Poll**

Practical **7.2**   **Political Opinion Poll**

You can simulate a political opinion poll in which the true proportion of Alliance voters is 30%. The following program (in BBC BASIC) will draw 1000 random numbers in the range 0–1. It will then find the proportion of random numbers less than 0.3 in that sample of 1000. This will give you the number of Alliance voters in that sample.

The program will repeat this process for 100 such samples and will tell you at the end how many of them gave sample proportions between 0.29 and 0.31, i.e. within 0.01 of the true population proportion.

```
10CLS
20S=0
30FOR I = 1 TO 100
40C=0
50FOR K=1 TO 1000
60R=RND(1)
70IF R < 0.3 C=C + 1
80NEXT K
901F C > = 290 AND C < 310 S=S + 1
100 PRINTTAB(5,8)C,S,I
110NEXT I
120PRINT"Proportion of samples size 1000 with proportion between 0.29 and 0.31 is"; S/100
```

C counts the number of random numbers less than 0.3, and S counts the necessary samples with proportion between 0.29 and 0.31.

To change the sample size, change the 1000 in line 50 and line 120 and the numbers in line 90; to change the number of samples, change the 100 in line 30 and line 120; to change the true value of p, change the 0.3 in line 70 and the numbers in lines 90 and 120. You might like to experiment to see what happens with different values of p and different sample sizes.

Section **7.5**   **Interval Estimates**

It can be seen that the main factor influencing the accuracy of the estimate is the sample size (or its square root).

Our opinion pollster may well not be satisfied with a process in which only about half the samples that he or she takes give the proportion of Alliance voters to within 1% of the true value and may well wish the figure to be nearer 0.95 (say) than 0.5. This idea is quite difficult to understand. It means that, if the pollster conducted 100 such surveys, on average 95 of them could be expected to give estimates which were accurate to within 1% of the true value. About 5 polls (roughly) would be expected to be more than 1% out in their estimates.

In order to increase the probability of being this accurate, the size of the sample will have to be increased.

For an area under the standard normal curve to be 0.95, we require z to lie between ± 1.96. To make calculations easier to follow, we shall increase the probability slightly above 0.95 and let z take the values ± 2 (Figure 7.14).

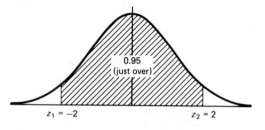

Figure 7.14 $N(0, 1)$

With an error of $1/100$ each side, using z_2,

$$2 = \frac{\sqrt{n} \times 1/100}{\sqrt{pq}}$$

$$\frac{2\sqrt{pq}}{1/100} = \sqrt{n}$$

$$\frac{4pq}{1/10\,000} = n$$

$$40\,000pq = n$$

With any value of error in the proportion of e each side, using z_2,

$$2 = \frac{e\sqrt{n}}{\sqrt{pq}}$$

$$\frac{2\sqrt{pq}}{e} = \sqrt{n}$$

$$\frac{4pq}{e^2} = n$$

The true values of p and q are still unknown but the maximum value for pq is $1/4$. So the maximum sample size required with an error of $1/100$ each side can be calculated as

$$40\,000pq = n$$

$$40\,000 \times \frac{1}{4} = n$$

$$10\,000 = n$$

and that with any value of error in the proportion of e each side,

$$\frac{4pq}{e^2} = n$$

$$\frac{4 \times 1/4}{e^2} = n$$

$$\frac{1}{e^2} = n$$

In general, if the size of the error has been decided, the maximum sample size can be calculated as $1/e^2$ for about 95% of the samples to have the proportion within the required range.

The amount of error that we can expect is proportional to the size of the sample (and not the size of the original population); so a poll based on a sample of 1000 voters throughout Great Britain should give a result with the same limits of accuracy as a poll of 1000 voters in a single constituency by-election. Although the second population is smaller, it is not possible to reduce the size of the sample without increasing the probability of error.

As we are sampling without replacement, the sample variance will be reduced. The magnitude of this effect is $(N - n)/(N - 1)$ (where N is the size of the population and n is the size of the sample). Even for a single constituency poll the effect of this factor will be small. If you refer back to Chapter 6, you will find this explained in more detail.

If our opinion pollster interviews 10 000 voters and finds that 4500 say that they will vote for the Alliance in the next election, he or she knows that the sample is large enough to be within 0.01 of the true proportion in the population (95% of the time); thus

$$n = 10\,000$$

$$A = 4500$$

The sample estimate for p is 0.45.

The pollster can give an interval of 1% either side within which it is believed that p (the true proportion) should lie, i.e. $0.44 < p < 0.46$. As the calculations are based on including 0.95 of the (normal) distribution of all possible results, this range is called a 95% confidence interval for p. We expect that in 95 polls out of 100 we should 'capture' the true value of p in our interval.

As we have used the maximum value for pq ($= 1/4$), 1% is an overestimate of the error that we should encounter. (Alternatively, we should encounter an error of more than 1% less than 95% of the time.)

If the opinion pollster interviewed 1000 people instead and found that 450 profess to be Alliance voters, the estimate of p is still 0.45. However, what will happen to the accuracy of the estimate?

We know that $n = 1/e^2$ and $n = 1000$; so

$$1000 = \frac{1}{e^2}$$

$$\frac{1}{1000} = e^2$$

$$0.001 = e^2$$

$$0.316 = e$$

The margin of error has now been increased to just over 3%. Thus the confidence interval is now 3% either side of 0.45, i.e. from 0.42 to 0.48 approximately.

This means that if the opinion pollster took a lot of samples of size 1000 and calculated the confidence intervals in this way, then 95% of these intervals would contain the true value of the population proportion. We do not know whether the particular interval includes the true value and we cannot say anything about the probability that one particular interval contains the true value. This is the difference between a confidence interval and a probability interval. Note that the margin of error is increased by the square root of any reduction in sample size; for instance, a twofold increase in error will be introduced by reducing our sample size by $1/4$.

Practical Follow-up 7.2

1. Assume that the true value of p is unknown in the population and that you must use the maximum value of $pq = 1/4$. Calculate the limits for the sample proportion based on a sample of 100, which can be expected with at least a probability of 0.95.

    If your estimate for p is much less (or much more) than 0.5, you can now see how the confidence interval is changed by using your estimated values of p and q.

    As we have previously used the maximum value for pq in calculating confidence intervals, the use of any other value of p will reduce the width of any such interval.

    The number of Alliance voters found by all possible opinion polls from that population should follow a normal distribution $A \sim N(np, npq)$ (Figure 7.15).

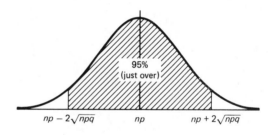

Figure 7.15 $A \sim N(np, npq)$

This interval either side of np includes A, the number of Alliance voters (or yes replies), with a probability of just over 0.95. It also follows that, if A is within a distance $2\sqrt{npq}$ of np, then np is within a distance $2\sqrt{npq}$ of A. The interval $A \pm 2\sqrt{npq}$ should include the true value of np for 95% of polls.

A 95% confidence interval for p requires us to divide throughout by n to give

$$\frac{A}{n} \pm \frac{2\sqrt{pq}}{n}$$

2. Use this method to calculate a 95% confidence interval for the true proportion of yes voters in the population from which your sample was drawn.

    While you now have a narrower confidence interval, you have used an estimate for the true values of p and q when you made your calculations. Do you consider this to be satisfactory or do you think it is better to be 'on the safe side' and work with the maximum value of $pq = 1/4$?

Section **7.6** **Stratified Sampling**

Practical 7.3   Television Viewing

Refer back to Chapter 6 and find Practical 6.4. It is based on the survey results among 1000 school pupils of their favourite television viewing. If you have previously worked through that practical, you will be able to adapt your results for this one.

1. Use random numbers to choose a random sample of 100 pupils from the entire school.
2. From your sample, estimate the proportion of pupils in the school choosing category B.
3. Give a 95% confidence interval for the proportion choosing category B.
4. Count the number of Bs for the whole school; then compare your estimates with the actual proportion who did choose category B.

Chapter 8 Estimation: Sampling Distributions and Point Estimates

Section **8.1** **Which Estimator?**

In 1985 a doctor in general practice who was interested in preventative medicine decided to take a random sample of 10% of his adult male patients. From the information gathered, he hoped to find some indications of the level of general health and fitness among the male patients in his practice. He also wished to investigate the possibility that certain factors might be linked, e.g. obesity and high blood pressure, or obesity and high levels of cholesterol in the blood and so on.

He selected his sample; then, stressing the benefits of a thorough check-up (and the furtherance of scientific knowledge, etc.), he asked each to attend an early morning appointment, at which a battery of tests and measurements would be taken.

Using the measurements taken from his sample, he wished to estimate, for example, the mean level of cholesterol in the blood for the entire population of men attending his practice, so that individuals can be compared with that mean.

In order to formulate his theories and hypotheses about relationships between measurements, it helps to have a theoretical model in mind for the distribution of each measure. For example, he can be almost certain that the heights of his male patients will form a distribution which can be modelled by the normal distribution. He may suspect then that blood cholesterol levels can also be modelled by a normal distribution (with a different mean and variance). If he does believe the normal distribution to be a useful model for a particular measure, he might adopt any of a number of estimators for estimating the mean of the population.

An estimator is a variable used to estimate a population parameter such as μ. In general, an estimator can be remembered as a method such as

$$\bar{x} = \sum_{i=1}^{n} \frac{x_i}{n}$$

while an estimate is a particular value found by applying that method to the values in a particular sample.

Model of the Population Measures

Estimator 1

As the model (Figure 8.1) for the population distribution is symmetrical and the median coincides with μ, he might use the sample median as an estimator for μ. This has the advantage of being easy to calculate and could usually still be found if one or two of the sample values were, say, illegible.

Estimator 2

Again using the symmetry of the model for the population distribution, the doctor may decide that the midrange of the sample may be used as an estimator for μ.

The sample midrange is the average of the highest and lowest values in that sample. While the population theoretically ranges from $-\infty$ to $+\infty$, it is symmetrical about μ

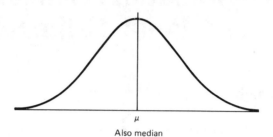

Figure 8.1 Model of the population measures: $N(\mu, \sigma^2)$

which is at the middle of the distribution. We shall investigate later how close the sample midrange is to μ.

Estimator 3

Finally, the third quantity proposed as an estimator for μ is the mean $\bar{x}$ of the sample.

As we go through this chapter, we try to judge which of these estimators the doctor should prefer. We also look at the problems of estimating the population variance from a sample.

Practical **8.1** **Blood Cholesterol Levels**

The blood cholesterol levels measured for the doctor's 537 patients are listed in ten columns across the table on page PC10. We use this group of patients as our population from which small samples can be taken so that we can see what forces are at work in the sampling process. The data are provided as a photocopiable sheet.

Each student is to select five samples each of 10 patients and also a further five samples each of five patients. The samples of different sizes should be recorded separately so that you will each require two copies of the special sheets provided as a photocopiable page (page PC11).

In a large group the number of samples collected by each student may be decreased as long as 50–100 samples of each size are collected by the class as a whole.

You need to take random samples and you may devise your own method using random numbers. If the class is using a table of random numbers, ensure that you all use different parts of the table; otherwise, you will end up with identical samples. A calculator with the statistical functions $\bar{x}, \sigma_n$ and σ_{n-1} will speed up the necessary calculations.

As we are taking simple random samples with replacement (see Chapter 6), it is permissible for the same person to be included more than once in any particular sample, should a particular random number occur more than once.

On your record sheet, write down each measurement as it is selected for the sample. Find and record the median of each sample. Then calculate the midrange which is the point midway between the largest and smallest values in that sample.

Use your calculator to find $\bar{x}$ but, when you calculate the sample standard deviation denoted by σ_n, square it before you record it on your sheet. This will give the sample variance with divisor n calculated thus:

$$\text{sample variance (divisor } n) = \frac{\sum(x - \bar{x})^2}{n}$$

or

$$\frac{\sum x^2}{n} - \left(\frac{\sum x}{n}\right)^2$$

For the last column, square the value given by your calculator as σ_{n-1} so that you have the sample variance with divisor $n-1$ calculated as

$$s^2 = \text{sample variance (divisor } n-1) = \frac{\sum(x-\bar{x})^2}{n-1}$$

or

$$\frac{\sum x^2}{n-1} - \frac{\sum x}{n-1}\frac{\sum x}{n}$$

Later in this chapter, we discuss why these two different versions of the variance exist and their relative merits as estimators for the population variance.

Section 8.2   Our Results

The weights of the doctor's 537 patients are listed at the end of this chapter (Table 8.7). Your practical sampling exercise used the measurements of their blood cholesterol levels. The data used below are from a similar exercise using the patients' weights (or more precisely their masses) measured in kilograms. (In common speech we talk about weighing people and finding their weight in kilograms. Technically this is their mass. In this text we may loosely talk about a person's weight when we mean their mass. Tables and diagrams refer to their mass in kilograms (kg).) In the following sections, you will be able to compare the results of your study with the results of ours and to see whether there are any underlying patterns and similarities.

You will notice that the samples that you have taken have produced different values for the sample median (or mean, or midrange, etc.). These different values taken by our estimators are called sample estimates. Figure 8.2 is a histogram showing the values of the median weight obtained from our 100 samples each of 5 patients.

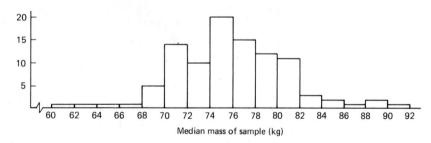

Figure 8.2   Distribution of median masses for samples of size $n = 5$

The distribution of *all* the medians from *all* possible samples of size 5 which could be taken from that population is called the sampling distribution of the median.

Before we look at the sample medians, midranges and means and try to choose which is the best estimator for the mean, we should define what we mean by 'best'. What properties would we like an estimator to have?

While each sample may produce a different estimate, we would like to think that, in the long run, the estimator that we have chosen as 'best' would produce the correct answer. That is, the average value of our estimates taken over a large series of samples should equal the population value:

E(sample estimator) = population parameter

An estimator which fulfils this condition is called an *unbiased estimator*.

As we are trying to find an estimate for μ the population mean, we can compare the estimators that we have used by looking at their average values over 100 samples. From the results in Table 8.1, there does not seem to be much to choose between our estimators. They have all produced mean values close to 76.10 kg (with one possible exception at 78.39 kg).

Table 8.1   Comparison of estimators for the mean mass μ of patients (population mean, 76.10 kg)

Estimator	Mean value of estimator (kg)	
	Samples of size $n = 5$	Samples of size $n = 10$
Sample median	75.38	75.35
Sample midrange	76.39	78.39
Sample mean	75.78	76.26

Given that the mean cholesterol level for the population of 537 patients is 1.64, make a similar comparison for your sample estimators. In fact, for a symmetrical distribution all three of the estimators used above are unbiased.

It seems that, in order to choose the best estimator for the population mean, we shall perhaps have to use some other criteria besides unbiasedness. We might like to specify that we would like an estimator whose sampling distribution is most closely clustered about the true value of the population parameter. That is, we would like to have the estimator whose sampling distribution has the smallest variance. This is called the most *efficient* estimator.

The distributions that we found for the sample midranges and means of our 100 samples of size 5 are shown below in Figures 8.3 and 8.4, respectively. Compare them with Figure 8.2 which shows the distribution of sample medians for that size of sample.

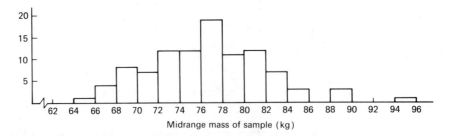

Figure 8.3   Distribution of midrange masses for samples of size $n = 5$

The variances of these distributions are given in Table 8.2. Of our three estimators, the sample mean appears to be the most efficient as it has the smallest variance.

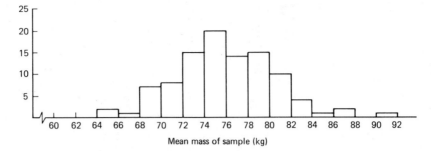

Figure 8.4 Distribution of mean masses for samples of size $n = 5$

Table 8.2 Variances for samples of size 5

	Variance for samples of size 5 (kg^2)
Sample medians	27.53
Sample midranges	29.74
Sample means	21.38

We may also feel that our estimator should give us a more accurate estimate, should we decide to take a larger sample. Certainly, we found this to be true (and very useful) when we used a sample proportion to estimate a population proportion (see Chapter 7). This implies that the estimates produced by large samples should be less variable than the estimates from small samples, i.e. the variance of the sampling distribution should decrease as the sample size increases. As we have taken samples of size 10 also, we can compare their sampling distributions with those for the samples of size 5.

Figures 8.5, 8.6 and 8.7 show the sampling distributions of the medians, midranges and means for samples of size 10. Compare each one with the equivalent distribution for samples of size 5.

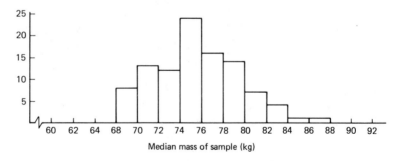

Figure 8.5 Distribution of median masses for samples of size $n = 10$

The sampling distributions of both the median and the mean appear less widely spread for the samples of size 10 and appear to be tending towards a more symmetrical peaked distribution. When the variances of all the sampling distributions are calculated (Table 8.3), the results give further weight to the impressions gained from examining the histograms.

While the variances of both the sample medians and the sample means are reduced by increasing the size of the sample, the variance of the sample mean is always smaller. Note

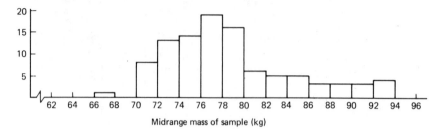

Figure 8.6 Distribution of midrange masses for samples of size $n = 10$

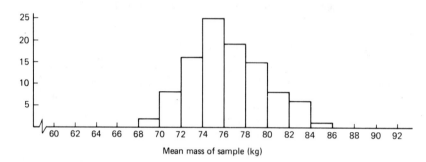

Figure 8.7 Distribution of mean masses for samples of size $n = 10$

Table 8.3 Variances for samples of size 5 and 10

	Variance (kg^2)	
	Samples of size 5	Samples of size 10
Sample medians	27.53	14.81
Sample midranges	29.74	34.89
Sample means	21.38	11.62

that by doubling the sample size we have reduced the variance of the sample means (and also the sample medians) by half. It is interesting also to compare the variance of the sample means with the original population variance which is 121:

population variance   121 kg^2
variance of $\bar{x}$ ($n = 5$)   21.38 kg^2
variance of $\bar{x}$ ($n = 10$)   11.62 kg^2

The variance of the sampling distribution of sample means appears to be decreasing in direct proportion to the sample size, i.e. samples of size 5 give a sampling distribution of means whose variance is one-fifth of the original population variance. Samples of size 10 give a sampling distribution of means with a variance $1/10$ of the original population variance. It appears that

$$\text{Var}(\bar{X}) = \frac{\sigma^2}{n}$$

where n is the sample size. A proof is given at the end of the chapter.

Also, while the variance of the median is also reduced in direct proportion of any increase in the sample size, it will also be larger than the variance of $\bar{X}$, in this case by a factor of $25/16$.

A *consistent* estimator is unbiased, i.e.

$E(\text{sample estimator}) = \text{population parameter}$

and it has a variance which decreases as the sample size increases so that, as $n \to \infty$, variance of sample estimate $\to 0$. While the sample mean and sample median are both consistent estimators of μ, the sample mean is more efficient as its sampling distribution always has a smaller variance.

Section 8.3   Analysis of Your Samples

***Practical Follow-up* 8.1**

1. Using the same class intervals for each, draw the sampling distributions of the following:

   a. Sample medians.
   b. Sample midranges.
   c. Sample means.

   Keep your two sizes of samples quite separate so that you have six distributions in all. Page PC12 can be used as a summary sheet for collecting together the results from everyone in the class.
2. Calculate the mean and variance for each of your six distributions.
3. Did your three estimators of μ appear to be unbiased estimators (given that $\mu = 1.64$)?
4. What was the most efficient estimator?
5. Did your results for $\text{Var}(\bar{x})$ confirm those of our study, i.e. did you find that

$$\text{Var}(\bar{x}) \approx \frac{\text{population variance}}{\text{sample size}}$$

The variance of your population is 1.00.
6. Which of your estimators appeared to be consistent estimators?
7. For the class samples calculate the following.

   a. Mean of medians.
   b. Variance of medians.
   c. Mean of midranges.
   d. Variance of midranges.
   e. Mean of means.
   f. Variance of means.

Section 8.4   Estimating the Variance

Figures 8.8 and 8.9 show the distributions of 100 sample variances for our two sample sizes (5 and 10). The variances have been calculated using a divisor of n, i.e. using the formula

$$\frac{\sum(x - \bar{x})^2}{n} \quad \text{or} \quad \frac{\sum x^2}{n} - \left(\frac{\sum x}{n}\right)^2$$

These results are for the weights of patients in the samples.

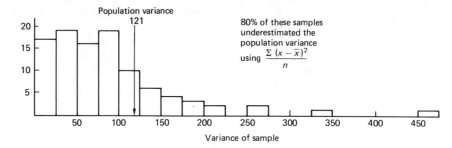

Figure 8.8   Distribution of sample variances for samples of size $n = 5$

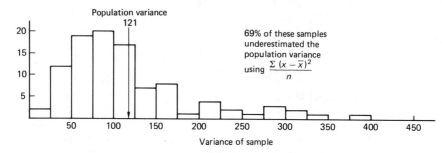

Figure 8.9   Distribution of sample variances for samples of size $n = 10$

As the population (distribution) variance for weights (i.e. masses) is 121 kg^2, it can be seen that the majority of sample variances underestimate this value. While the distribution of sample means appears to be approaching a symmetrical distribution grouped about the population parameter, this does not seem to be true for sample variances.

The means of the sample variances (found by squaring the result of the σ_n key on a calculator) for these groupings of 100 samples are given in Table 8.4.

Table 8.4   Means of the sample variances

	Samples of size $n = 5$	Samples of size $n = 10$
Mean of sample variance	85.5	115.45

The variances calculated for the smaller samples underestimate the population variance most markedly, but even for samples of size 10 the sample variance is a *biased* estimator:

$$E\left(\frac{\sum (X - \bar{X})^2}{n}\right) \neq \text{population variance } \sigma^2$$

We have used

$$\sum_{i=1}^{n} \frac{(x_i - \bar{x})^2}{n}$$

to calculate the sample variance, whereas the true population (distribution) variance will be

$$\sum_{i=1}^{n} \frac{(x_i - \mu)^2}{n}$$

In a real sampling situation, we do not know the value of μ and so have to use $\bar{x}$ as an estimate for μ. It is this factor which introduces bias into our estimate. For any group of numbers the mean $\bar{x}$ is calculated so that the sum of the deviations from $\bar{x}$ equals zero.

$$\sum_{i=1}^{n} (x_i - \bar{x}) = \sum_{i=1}^{n} x_i - \sum_{i=1}^{n} \bar{x}$$

If these deviations are squared, their total will be a minimum and therefore less than the squared deviations taken from μ (or any other arbitrary value). A proof is given at the end of this chapter.

You may demonstrate this with any one of the samples that you have taken. Here is a sample of 5 patients' weights:

75 kg   85 kg   68 kg   72 kg   64 kg

For this sample the mean $\bar{x} = 72.8$, and the population mean $\mu = 76.1$. The deviations from the mean can be calculated (Table 8.5) and it can be seen that

$$\sum_{i=1}^{5} (x_i - \bar{x}) = 0$$

and

$$\sum (x - \bar{x})^2 = 254.8 < \sum (x - \mu)^2 = 309.25$$

Table 8.5   Calculations of $\Sigma(x - \bar{x})^2$ and $\Sigma(x - \mu)^2$

x	$x - \bar{x}$	$(x - \bar{x})^2$	$x - \mu$	$(x - \mu)^2$
75	2.2	4.84	−1.1	1.21
85	12.2	148.84	8.9	79.21
68	−4.8	23.04	−8.1	65.61
72	−0.8	0.64	−1.41	16.81
64	−8.8	77.44	−12.1	146.41
Σ	0.0	254.80	−16.5	309.25

Section 8.5   An Unbiased Estimator of the Population Variance

The population variance is defined as

$$\sigma^2 = E\left(\frac{\sum_{i=1}^{n} (X_i - \mu)^2}{n} \right)$$

The differences from μ can be split into two parts so that

$$\sigma^2 = E\left(\frac{\sum_{i=1}^{n} [(X_i - \bar{X}) + (\bar{X} - \mu)]^2}{n} \right)$$

Expanding the square gives

$$\sigma^2 = E\left(\frac{\sum_{i=1}^{n}[(X_i - \bar{X})^2 + 2(X_i - \bar{X})(\bar{X} - \mu) + (\bar{X} - \mu)^2]}{n}\right)$$

As

$$E(X_i - \bar{X}) = 0 \text{ for all } i$$

the middle term equals 0 in the numerator above. The last term

$$E\left(\frac{1}{n}\sum_{i=1}^{n}(\bar{X} - \mu)^2\right) = E[(\bar{X} - \mu)^2]$$

is the variance of $\bar{X}$ shown in proofs at the end of the chapter to be σ^2/n. So

$$\sigma^2 = E\left(\frac{\sum_{i=1}^{n}(X_i - \bar{X})^2}{n}\right) + \frac{\sigma^2}{n}$$

$$\sigma^2 - \frac{\sigma^2}{n} = E\left(\frac{\sum_{i=1}^{n}(X_i - \bar{X})^2}{n}\right)$$

$$\sigma^2\frac{n-1}{n} = E\left(\frac{\sum_{i=1}^{n}(X_i - \bar{X})^2}{n}\right)$$

Thus

$$\sigma^2 = \frac{n}{n-1}E\left(\frac{\sum_{i=1}^{n}(X_i - \bar{X})^2}{n}\right)$$

$$\sigma^2 = E\left(\frac{n}{n-1}\frac{\sum_{i=1}^{n}(X_i - \bar{X})^2}{n}\right)$$

$$\sigma^2 = E\left(\frac{\sum_{i=1}^{n}(X_i - \bar{X})^2}{n-1}\right)$$

So an unbiased estimate of the variance is

$$s^2 = \frac{\sum_{i=1}^{n}(x_i - \bar{x})^2}{n-1}$$

If S is taken as the random variable whose values are s, $E(S^2) = \sigma^2$. It is this formula which is used when you square the result of the σ_{n-1} key on your calculator.

For our sampling study on patients' weights, we found that using this unbiased estimator for the variance gave better results when the means of all the sample variances were calculated: for $n = 5$, mean of $s^2 = 106.9$ kg^2

and, for $n = 10$,

mean of $s^2 = 128.3$ kg^2

These mean values compare much more favourably with the population variance of 121 kg^2.

Section 8.6 Further Analysis of Your Samples

Practical Follow-up 8.2

1. First consider the sample variance calculated by

$$\frac{\sum(x - \bar{x})^2}{n}$$

(or $(\sigma_n)^2$ on your calculator). Calculate the means of all your sample variances for samples of sizes 5 and 10 separately. Compare your results with the population variance of 1.00.
2. Using any one of your samples, verify that

$$\sum_{i=1}^{n} (x_i - \bar{x})^2 < \sum_{i=1}^{n} (x_i - \mu)^2$$

where $\mu = 1.64$ for your population.
3. Using the unbiased estimates for variance

$$s^2 = \frac{\sum_{i=1}^{n} (x_i - \bar{x})^2}{n - 1}$$

(or $(\sigma_{n-1})^2$ on your calculator) find the mean of these sample variances. Compare this value with the population variance.

Section 8.7 The Distribution of $\bar{X}$: Finding a Model

The distribution of weights of the 537 male patients is shown in Figure 8.10. While it might possibly be modelled by a normal distribution, there does seem to be a long tail to the right. There are a few very heavy people in the group.

The mean and variance of this population are 76.1 kg and 121 kg^2, respectively.

If the means of our 100 samples of size $n = 10$ are considered (Figure 8.11), a more compact distribution is formed, and the long tail to the right is less marked. The mean of this set of 100 sample means is 76.26 kg (very close to μ) and the variance is 11.62 kg^2.

The mean of 10 patients' weights has only one-tenth the variability of the original population from which the samples are selected. This happens because, within each sample, very heavy and very light patients tend to cancel each other out when their average is taken.

If we examine this distribution drawn with smaller class intervals as in Figure 8.7, we can see the appearance of a smoother shape.

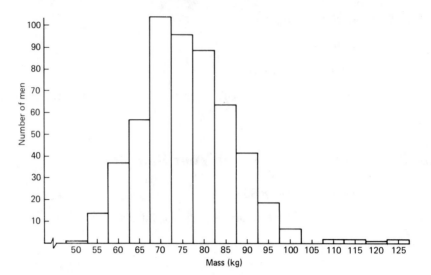

Figure 8.10 Masses of 537 men

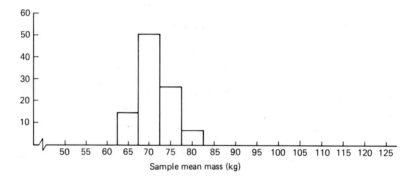

Figure 8.11 Mean masses of 100 samples of size $n = 10$

In fact, we would find that for any large set of samples the distribution of sample means forms a symmetrical peaked distribution with the highest frequencies for values of $\bar{x}$ close to μ. This effect becomes even more marked if larger-sized samples are taken.

If we had time to take *more* samples of size 10, perhaps 1000 or even 10 000, we could have used smaller class intervals for our frequency distribution of sample means. In that case the resulting histogram would become smoother as the number of samples increased and the underlying distribution could be modelled by a normal curve.

The sample means will form a normal distribution with mean μ and variance σ^2/n (Figure 8.12).

If all possible samples are considered, the mean weights of samples of size $n = 10$ taken from the population of male patients modelled as $N(76.1, 121)$ form a normal distribution as in Figure 8.13.

If samples of size $n = 5$ are considered, the distribution of all possible sample mean weights form a normal distribution also (Figure 8.14).

The standard deviation of the distribution of $\bar{X}$ calculated as $\sigma/\sqrt{n}$ is called the *standard error* of the mean.

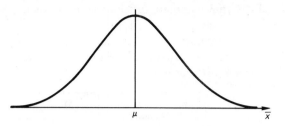

Figure 8.12 $\bar{X} \sim N(\mu, \sigma^2/n)$

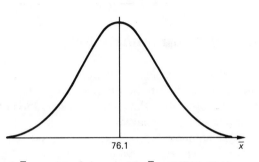

Figure 8.13   Distribution of $\bar{X}$ samples of size $n = 10$: $\bar{X} \sim N(76.1, 12.1)$

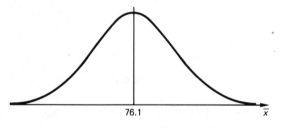

Figure 8.14   Distribution of $\bar{X}$ samples of size $n = 5$: $\bar{X} \sim N(76.1, 24.2)$

Practical Follow-up **8.3**

The distribution of blood cholesterol measurements for the doctors' 537 male patients is given below in Table 8.6.

Note that this distribution is not normal. The population mean for this group is 1.64 and the variance $\sigma^2 = 1.00$.

Table 8.6   Distribution of blood cholesterol measurements

Middle of interval	Number of observations
0	127
2	351
4	48
6	11
8	0

Find your sampling distribution of $\bar{X}$ for $n = 5$ and $n = 10$. You should find that your results are close to the following.

Your distribution of $\bar{X}$ for samples of size 10 should be a more compact, more symmetrical distribution than that of $\bar{X}$ for samples of size 5 and should be approaching the shape of a normal distribution.

Sample size	Mean of $\bar{x}$	Variance of $\bar{x}$
5	1.64	0.2
10	1.64	0.1

Section 8.8   An Interval Estimate for μ

If we know the mean and variance of the original population, it is now possible to derive the sampling distribution of μ for any sample size. Irrespective of the distribution of the original population, the sampling distribution of $\bar{x}$ will form a normal distribution provided that n is sufficiently large.

If the standard normal curve is considered, we know that 95% of the area under the curve lies between -1.96 and $+1.96$ (Figure 8.15).

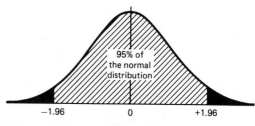

Figure 8.15   Normal distribution $N(0,1)$: ■, $2\frac{1}{2}$% of the area lies at each end of the distribution outside the range $-1.96 \leqslant Z \leqslant 1.96$

We also know that, if a large number of samples were taken, about 95% of the sample means would lie within 1.96 standard deviations (called standard errors in this case) of the population mean (Figure 8.16).

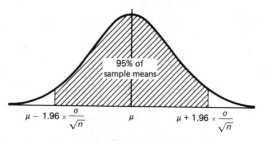

Figure 8.16   Distribution of sample means: $\bar{X} \sim N(\mu, \sigma^2/n)$

If we consider the sampling distribution of X for samples of size 10 from the population of male patients' weights, we find 95% of all possible sample means lie within this range (Figure 8.17).

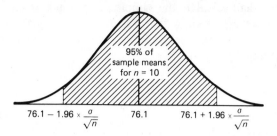

Figure 8.17   Distribution of sample means: $\bar{X} \sim N(76.1, 12.1)$

For $n = 10$,

$$\text{standard error} = \frac{\sigma}{\sqrt{n}}$$

$$= \frac{11}{\sqrt{10}}$$

$$= 3.479$$

95% of all sample means should lie between

$$\mu - 1.96 \times \frac{\sigma}{\sqrt{n}} \quad \text{and} \quad \mu + 1.96 \times \frac{\sigma}{\sqrt{n}}$$

$76.1 - 1.96 \times 3.479$ and $76.1 + 1.96 \times 3.479$

$76.1 - 6.819$ and $76.1 + 6.819$

i.e.

69.281 and 82.919

We can calculate a probability interval for $\bar{x}$ if we know the values of μ and σ^2 (Figure 8.18).

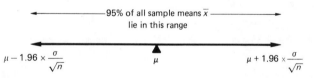

Figure 8.18   Probability interval for $\bar{x}$

However, in reality, we are unlikely to know the true value of μ, and what we would like to do is to calculate an interval based on $\bar{x}$, our sample estimate, which we hope will contain μ. Refer back to Chapter 7 to see how a confidence interval may be constructed.

A Confidence Interval Based on One Sample

In a typical situation the statistician takes one random sample and tries to estimate μ from this sample.

Here is a random sample of 10 male patients' weights (i.e. masses):

72 kg   79 kg   80 kg   80 kg   68 kg   90 kg   59 kg   88 kg   86 kg   70 kg

The best point estimate that we have for μ is $\bar{x} = 76.9$. (Assuming that we do not know μ, we shall calculate the sample mean which is an unbiased and consistent estimate for μ.) However, we know that μ is unlikely to equal 76.9 exactly.

We can construct a confidence interval which we hope will contain μ. A 95% confidence interval is

$$\bar{x} \pm 1.96 \times \text{standard error}$$

where the standard error is $\sqrt{\sigma^2/n}$.

We are in the unusual situation here of knowing the value of the population variance $\sigma^2 = 121$. So in fact, we can use $\sqrt{\sigma^2/n} = 3.47$. (If σ^2 is not known, then s^2, the unbiased estimate for the population variance, should be used. However, for small samples, the normal distribution will no longer apply and the t distribution is needed.)

A 95% Confidence Interval for μ

Using the population variance $\sigma^2 = 121$ (which is usually not known)

$$\bar{x} \pm 1.96 \frac{\sigma}{\sqrt{n}} = 76.9 \pm 1.96 \times 3.47$$

$$= 76.9 \pm 6.80$$

We hope that μ lies between 70.1 and 83.7. In fact, the true value of $\mu = 76.1$ (we usually do not know this value) does lie in our interval.

Confidence Intervals of Different Sizes

By consulting standard normal distribution tables, we find that 99% of the $N(0,1)$ distribution lies between -2.33 and $+2.33$. We can use this to construct a 99% confidence interval for μ, as

$$\bar{x} \pm 2.33 \frac{\sigma}{\sqrt{n}}$$

This is a wider interval than the 95% confidence interval, but we expect only one out of every 100 samples to produce a 99% confidence interval which does not include μ.

Similarly, we may construct a confidence interval of any size.

Practical Follow-up 8.4

1. Construct a 95% confidence interval for μ, the mean blood cholesterol measure, using one of your samples size $n = 10$.
2. Compare this with a confidence interval based on a sample size $n = 5$.
3. If you collect together a series of 95% confidence intervals (from the same-sized samples) among the students in your class, you should find that roughly 5 out of 100 confidence intervals (or 1 out of 20) do not in fact 'capture' the true value of μ.

Table 8.7   Masses of 537 male patients

Mass (kg)									
78	66	58	58	87	74	75	79	84	63
93	77	76	58	73	69	65	74	69	81
83	64	68	85	58	73	59	66	80	76
83	59	83	53	70	76	65	67	79	65
93	70	92	62	77	64	88	60	70	76
64	90	82	92	84	77	71	74	92	68
57	93	78	78	73	69	67	69	69	76
89	81	79	75	91	68	77	56	71	71
65	67	76	63	87	81	60	74	87	101
83	74	70	109	89	83	67	79	59	102
74	58	72	87	84	74	73	81	71	78
86	66	57	80	73	84	54	74	70	92
69	73	76	83	77	67	95	51	60	72
75	88	68	60	77	93	92	65	93	77
72	68	97	64	72	60	54	89	80	74
66	65	84	77	93	63	76	87	85	101
87	67	66	66	111	126	87	88	62	68
62	78	80	72	62	65	64	72	78	75
73	83	81	75	78	83	64	74	85	72
53	71	69	92	88	75	75	61	74	88
74	89	101	87	64	75	53	81	78	87
66	55	94	66	75	61	72	79	66	65
79	63	78	72	72	67	70	65	78	95
61	61	75	61	86	54	77	75	81	86
80	78	89	83	66	78	84	88	70	69
73	65	82	83	115	76	89	84	71	74
73	86	80	82	86	68	80	72	72	92
72	71	74	79	74	91	81	78	69	80
76	72	89	92	76	91	68	63	72	78
53	69	82	71	84	84	70	72	89	78
72	61	95	77	66	84	59	95	79	87
68	78	78	80	68	82	78	79	74	74
85	80	69	89	85	76	70	75	80	66
95	55	94	66	70	89	68	71	78	82
68	78	79	94	71	75	61	86	75	78
86	70	87	61	88	72	62	92	68	71
87	85	67	70	79	74	65	60	79	75
71	80	74	87	70	78	79	69	67	70
73	82	83	79	74	82	72	71	74	68
76	75	86	82	78	68	88	72	85	90
68	75	75	76	64	77	82	78	88	77
91	68	75	127	72	68	83	91	76	92
87	80	78	64	87	60	72	59	81	82
71	73	80	82	78	74	57	82	68	75
99	69	76	76	76	70	89	86	83	80
81	77	95	87	78	72	73	93	79	64
61	80	63	113	66	76	87	71	97	78
86	80	69	60	72	69	63	69	89	81
67	86	86	73	76	63	68	80	64	87
70	73	60	78	74	76	70	84	76	57
79	68	60	99	89	89	70	120	82	71
70	71	76	60	69	83	80	70	61	71
73	73	67	62	87	80	84	98	69	67
68	77	78	67	88	72	91			

The Sample Mean as an Unbiased Estimator for μ

To prove that

$$E(\bar{X}) = \mu$$

we require the result that for any X_1 and X_2,

$$E(aX_1 + bX_2) = aE(X_1) + bE(X_2)$$

Now

$$E(\bar{X}) = E\left(\frac{\sum_{i=1}^{n} X_i}{n}\right)$$

$$= \frac{1}{n} E(X_1 + X_2 + X_3 + \ldots + X_n)$$

$$= \frac{1}{n}[E(X_1) + E(X_2) + E(X_3) + \ldots + E(X_n)]$$

$$= \frac{1}{n}(\mu + \mu + \mu + \ldots + \mu)$$

$$= \frac{1}{n}(n\mu)$$

$$= \mu$$

Variance of the Sampling Distribution of $\bar{X}$

To prove that

$$\text{Var}[\bar{X}] = \frac{\sigma^2}{n}$$

we require the result that

$$\text{Var}(aX_1 + bX_2) = a^2\,\text{Var}(X_1) + b^2\,\text{Var}(X_2)$$

so that

$$\text{Var}(\bar{X}) = \text{Var}\left(\frac{\sum_{i=1}^{n} X_i}{n}\right)$$

$$= \frac{1}{n^2}[\text{Var}(X_1) + \text{Var}(X_2) + \text{Var}(X_3) + \ldots + \text{Var}(X_n)]$$

$$= \frac{1}{n^2}(\sigma^2 + \sigma^2 + \sigma^2 + \ldots + \sigma^2)$$

$$= \frac{1}{n^2}(n\sigma^2)$$

$$= \frac{\sigma^2}{n}$$

Proof that, if $\sum_{i=1}^{n} (x_i - a)^2$ is Minimised for a Given Set of x_i, then a Must Equal $\bar{x}$

Let

$$L = \sum_{i=1}^{n} (x_i - a)^2$$

$$\frac{dL}{da} = 2\sum(x_i - a)$$

At a minimum

$$\frac{dL}{da} = 0$$

$$0 = 2\sum_{i=1}^{n} (x_i - a)$$

$$0 = \sum_{i=1}^{n} (x_i - a)$$

$$0 = \sum_{i=1}^{n} x_i - na$$

$$na = \sum_{i=1}^{n} x_i$$

$$a = \frac{\sum_{i=1}^{n} x_i}{n}$$

$$a = \bar{x}$$

Chapter 9 Hypothesis Testing I

Section **9.1**   **The First Problem: Identification of a Newspaper Article**

How fall-out clouds the political climate

Radiation from the Ukraine may, or may not, kill a few dozen Britons by the year 2036. We and our children will have to wait to discover. The political fall-out, however, could be felt a lot sooner.

In the wake of last week's election successes, Neil Kinnock would not be human—or ambitious—if he failed to exploit Chernobyl for all it is worth.

Sceptics will say that voters are not influenced by industrial accidents in obscure Soviet republics.

The polls suggest, however, that in a week when sales of bottled water and long-life milk soared, the British public saw their daily diet and radioactivity all too closely linked. Backing for the nuclear industry has slumped from 60 percent to 40 percent.

The public mood on British nuclear installations is anything but calm. Britain, people are saying, may be more safety-minded than the Soviet Union. The comfort to be derived, however, is reduced by the knowledge that we have two thirds as many reactors (37 to their 51), contained within a ninetieth of the physical space, and surrounded by a population 20 times as dense.

It is conceivable, of course, that nuclear worries will pass from the political atmosphere almost as fast as the carcinogenic particles at present dispersing over the North Atlantic. This is what Tory ministers are hoping. Environmental groups, however, will make sure that every step is monitored.

The beneficiaries will be the Opposition parties—both less nuclear than the Tories. The Alliance, however, is split between the greenish Liberals and the technocratic SDP, while Labour remains conscious of the 150,000 jobs in, or connected with, the industry.

There are signs that Labour's attitude may shift further in the direction of a phase-out—the line taken by last year's party conference. Even those unions with most at stake—the Municipal Workers and the electricians—have been disturbed by recent events.

Hitherto the word "nuclear", linked to "disarmament", has been an albatross for Labour. Now nuclear avoidance could become an international fashion and a selling-point.

A student working on a project entitled 'Energy' finds this newspaper article among his notes. He had evidently forgotten to note down the name of the paper or the date when it appeared. He thinks he can pinpoint the date fairly accurately (to within a few days) but the source of the article is more problematic.

It seems likely that he found it in one of the two daily newspapers delivered to his home *The Guardian* and *Today*.

He takes the article in to school and explains the problem to his teacher. She asks him to bring in the next day's copies of *Today* and *The Guardian* so that the class may undertake some statistical detective work. By investigating the characteristics of the articles in both newspapers, she hopes that they will be able to identify the source of the anonymous article.

Practical 9.1   **Characteristics of Newspapers**

Obtain copies of *Today* and *The Guardian* (any date will do) and look through them. What characteristics would you choose to discriminate between the two papers? Although their layout and general presentation are obviously different, we cannot use those features to help us to identify the source of the article reproduced at the beginning of this chapter.

Statistical methods have been used in the past to decide the authorship of 12 of the *Federalist* papers published in 1787. Two writers, Alexander Hamilton and James Madison, both claimed to have written the 12 articles in question. The recent analysis by two statisticians, Mosteller and Wallace, involved the use of 'marker' words, favoured by one writer but not the other. They found, for instance, that Madison used both 'while' and 'whilst' in his other writings but Hamilton never used 'whilst' in his. This enabled them to decide on the authorship of five of the disputed papers.

Unfortunately, for our analysis, many different journalists write for *The Guardian* and *Today*; so we cannot focus on personal idiosyncracies in the writing, such as the frequency with which particular words occur. We have to concentrate on general stylistic measures to distinguish between the two newspapers.

Here are some suggestions which you can consider, and also try to think of others of your own.

1. The lengths of the words used.
2. The lengths of the sentences.

You will need to obtain a measure for each of the characteristics that you have chosen, for both newspapers, so that the anonymous article may be compared with those results. It might just be possible for a large group of students to work through an entire copy of each paper (omitting advertisements, horoscopes, weather reports and cartoons, etc.). However, that task would be very time consuming, and the alternative is to select a large random sample of sentences from each.

Three sets of random numbers can be used to make each selection for the starting point of your sample.

a. Page (two-digit number).
b. Column (one-digit number).
c. Line (two-digit number).

When you have found your starting point, move on to the next complete sentence and use the next five or ten sentences for your sample. You should aim to collect 500 sentences or so in your total sample for each newspaper. You need a large enough sample to overcome individual fluctuations. For the purpose of analysis, we assume that these large samples *are* the underlying distributions.

Section 9.2   Newspaper Analysis

The data obtained from our samples from *Today* and *The Guardian* are given in this section. Use our results only if you are not able to analyse your own copies.

Draw line graphs of the distribution of word length from your samples of the two newspapers (Figures 9.1 and 9.2).

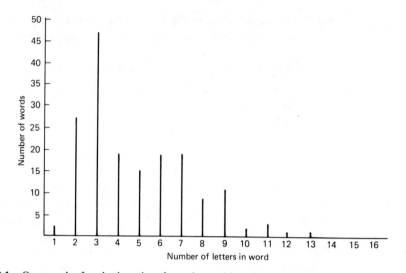

Figure 9.1   Our results for the lengths of words used in *The Guardian*: mean, 4.82 letters

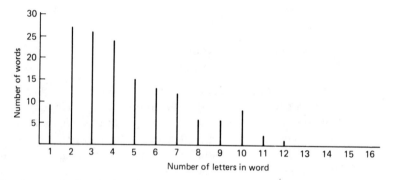

Figure 9.2   Our results for the lengths of words used in *Today*: mean, 4.63 letters

What do they show?

Compare your answers with the following conclusions that we have drawn from our samples.

While *The Guardian* appears to have a slight preponderence of 3-letter words, there are no other real differences in the two distributions. When the means and variances of the two distributions are calculated, the similarity becomes even more apparent (Table 9.1).

Table 9.1   Means and variances

	The Guardian	*Today*
Mean number of letters per word	4.82	4.63
Variance of word length	6.32	6.89

As the two distributions are so similar, it will be difficult to use an analysis of word length to identify the source of the student's article.

Fortunately, however, when the distributions of sentence lengths are similarly analysed, quite a noticeable difference emerges. *The Guardian* has many more long sentences than does *Today*.

Draw diagrams of the distribution of sentence length for your samples and calculate the mean and variance in each case (Figures 9.3 and 9.4).

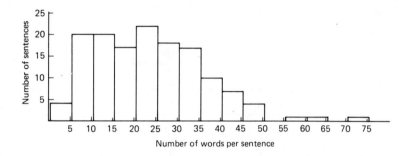

Figure 9.3 Our results for the sentence lengths for a sample from *The Guardian*

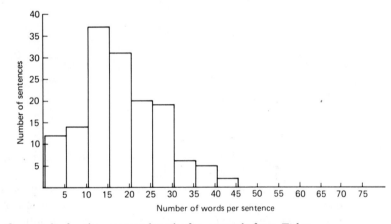

Figure 9.4 Our results for the sentence lengths for a sample from *Today*

The means and variances of our distributions of these are given below in Table 9.2.
The lengths of the sentences in the unidentified article will be more likely to help us to decide which newspaper it came from.

Table 9.2 Means and variances

	The Guardian	Today
Mean number of words per sentence	23.68	18.17
Variance of sentence length	165.89	79.03

Section 9.3 Identifying the Source

Now look at the article in question and compare the lengths of the sentences used in it with those used in *The Guardian* and *Today*.

Table 9.3   Distribution of sentence lengths in unidentified article

Number of words in sentence	Frequency
1–5	0
6–10	3
11–15	8
16–20	2
21–25	1
26–30	4
31–35	0
36–40	1
41–45	0
Total	19

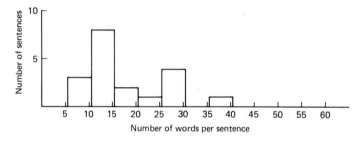

Figure 9.5   Lengths of sentences in unidentified article

The unidentified article has 19 sentences and the distribution of sentence lengths is found to be as in Table 9.3 and Figure 9.5.

As there are only 19 sentences, we do not have enough to form a smooth frequency distribution. From the diagram, it is not easy to tell whether the article comes from *The Guardian* or *Today* by comparing the frequency distributions for sentence lengths.

However, we can adopt a more logical approach to the problem. There are two possible hypotheses.

Hypothesis A   The article comes from *The Guardian.*
Hypothesis B   The article comes from *Today.*

At this stage, both hypotheses have equal weight and there is no reason to favour either hypothesis A or hypothesis B.

Comparing visually the actual frequency distribution of sentence length may indicate which paper the article came from or it may not. In any case, we want to make some calculations to assess the strength of the evidence.

To do this, we take some measurement based on the sample—a test statistic—which should be able to help us to distinguish between the two possible sources. One possible test statistic is the sample mean, i.e. the mean number $\bar{x}$ of words per sentence. Another is the proportion of sentences with more than 30 words. Can you think of any other possibilities?

We use $\bar{x}$ and, in order to make a judgement, we have to compare it with the sampling distributions which would be formed by the means of all possible samples of 19 sentences which could be taken from

a. *The Guardian* and
b. *Today.*

The sampling distribution of $\bar{X}$ will be approximately normal when all possible samples are considered. The variance of this sampling distribution is calculated by σ^2/n where n is the size of the sample (refer to Chapters 6 and 8). In order to conduct this analysis, we are taking our large samples of sentences as correctly representing the true underlying distributions of sentence lengths for the two newspapers.

The sampling distributions of the mean number of words in samples of 19 sentences in *The Guardian* and *Today* are shown in Figures 9.6 and 9.7, and also in Table 9.4.

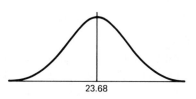

23.68

Figure 9.6   Sampling distribution $\bar{G} \sim N(23.68, 165.89/19)$ of the mean number of words in samples of 19 sentences taken from *The Guardian*

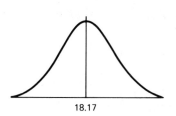

18.17

Figure 9.7   Sampling distribution $\bar{T} \sim N(18.17, 79.03/19)$ of the mean number of words in samples of 19 sentences taken from *Today*

Table 9.4   Sampling distributions of the mean number of words per sentence

	The Guardian	*Today*
Mean of distribution of means	23.68	18.17
Variance of distribution of means	$\dfrac{165.89}{19} = 8.73$	$\dfrac{79.03}{19} = 4.16$
Standard error $\sigma/\sqrt{n}$	$\sqrt{8.73} = 2.95$	$\sqrt{4.16} = 2.04$

Next we have to see how the mean sentence length of our unidentified article will fit in with either of the two sampling distributions. We examine the two hypotheses in turn.

The test statistic under consideration will be the mean length of the 19 sentences in our unidentified article. We can draw the two distributions of $\bar{X}$ under hypothesis A and hypothesis B on the same graph (Figure 9.8) and decide for particular values of $\bar{x}$ which hypothesis is likely to be correct.

$\bar{x} =$ mean length of 19 sentences

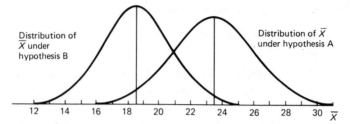

Distribution of $\bar{X}$ under hypothesis B

Distribution of $\bar{X}$ under hypothesis A

12   14   16   18   20   22   24   26   28   30 $\bar{X}$

Figure 9.8 Two distributions of $\bar{X}$, the mean length of 19 sentences

Clearly, the lower the value of $\bar{X}$, the more likely we are to decide that the passage came from the hypothesis B distribution rather than the hypothesis A distribution. So our decision takes the form

if $\bar{x} < a$ (some particular value to be decided)

then reject hypothesis A in favour of hypothesis B; else do not reject hypothesis A.

The region in which we are going to reject hypothesis A is called the critical region.

It is always possible that any (positive) value of $\bar{x}$ could come from hypothesis A; so there is a possibility of making a wrong decision. The probability that the value of $\bar{x}$ lies in the critical region when hypothesis A is true is called the size of the test, or the significance level of the test.

Suppose that we choose $a = 20$ then the critical region is

$P(\bar{x} < 20)$ when $\bar{X} \sim N(23.68, 8.73)$

$$= P\left(Z < \frac{-3.68}{2.95}\right)$$

$$= P(Z < -1.24) \text{ when } Z \sim N(0, 1)$$

$$= 0.1075$$

i.e. the significance level or size of this test is 10.75%.

With the same decision rule, there is a possibility of making a different type of error. We could have a mean sentence length greater than 20 from *Today*, and we shall have accepted hypothesis A using our decision rule whereas in fact hypothesis B is true. This is called a type 2 error. The probability that this happens is

$P(\bar{x} > 20 | \text{hypothesis B})$

$$= P(\bar{x} > 20) \text{ when } \bar{X} \sim N(18.17, 4.16)$$

$$= P\left(Z > \frac{20 - 18.17}{2.04}\right)$$

$$= P(Z > 0.897) \text{ where } Z \sim N(0, 1)$$

$$= 1 - 0.8152$$

$$= 0.1848$$

In fact, our mean value was 17.89, which is less than 20; so we infer that the article comes from *Today*. This is seen to be reasonable from its position on the diagram. Clearly, it is much more likely to have come from the distribution under hypothesis B than under hypothesis A.

Alternatively, if we choose our critical region so that 17.89 lies just inside it, then any value of $\bar{x} < 17.89$ would be in the critical region.

$$P(\bar{x} < 17.89 | \text{hypothesis A}) = P\left(Z < \frac{-5.79}{2.04}\right)$$
$$= P(Z < -2.838)$$

which is 0.0024 or 0.24%. This is called the p value of our result. It says that our particular result would have shown significance on a test of size 2.4%.

From the probabilities that we have calculated, we infer that the article does come from *Today*.

Section 9.4 A Second Article for Identification

Practical 9.2   Another Newspaper Article

Read the following article.

PC suspended after gun 'horseplay'

A policeman on armed anti-terrorism duties at Birmingham airport has been suspended after a woman colleague was hit in the leg by a firearms training pellet.

The West Midlands force, which has one of the worst firearms records in Britain, suspended the officer on Monday pending an inquiry. The incident has proved doubly embarrassing for the force because it involved the firing of a training pellet from a .38 Smith and Wesson handgun by an officer who was supposed to be on armed security duties in the event of a terrorist attack.

Neither the suspended officer, said to be in his twenties, or the woman officer has been named. The policewoman, also in her twenties, suffered bruising and reported unfit for duty after seeing a police surgeon.

It was thought that the incident occurred during locker room 'horseplay' when a number of officers were present.

Last month the West Midlands Police Authority firearms committee considered a report expressing concern at the 'shocking record of injuries suffered by innocent people' as a result of firearms being discharged in the region during the past six years.

In 1980, a 16-year-old pregnant girl, Gail Kinchin, was killed when she was hit by police bullets during a siege at her boy friend's flat in which she was used as a human shield. The boy friend was later convicted of her manslaughter.

In 1982, there were two incidents involving accidental discharges. In one, a shot narrowly missed two sleeping children.

Last year, five-year-old John Shorthouse was shot dead when armed police raided his Birmingham home. PC Brian Chester, of the West Midlands Tactical Firearms Unit, has been committed for trial, accused of his manslaughter.

Mr Les Sharp, the West Midlands Deputy Chief Constable, said yesterday that a senior officer had been appointed to investigate the airport incident. A report would be submitted to the Director of Public Prosecutions.

● A policeman who shot himself in the leg during weapons training is suing the Humberside force for personal injury damages. A spokesman for Humberside police said yesterday that an internal inquiry into the incident in which PC Robert Wellings, aged 31, of Hull, was wounded had been held, but as a civil action had been started the matter was *sub judice*.

Form a frequency distribution of the number of words per sentence.
Find the mean number of words per sentence.
The results found in Section 9.2 were as follows.

	The Guardian	Today
Mean number of words per sentence	23.68	18.17
Variance of sentence length	165.89	79.03

You may have conducted your own study in Practical 9.1, in which case you will have your own results to substitute for these.

Find and sketch the sampling distribution of $\bar{X}$, all possible sample means from all possible samples of size $n = 17$ taken from

a. *The Guardian* and
b. *Today*.

Taking the decision rule, if $\bar{x} < 20$, reject hypothesis A in favour of hypothesis B; else do not reject Hypothesis A.

Find the size of the critical region, i.e. $P(\bar{x} < 20:$ hypothesis A is true$)$.

Find the probability of a type 2 error, i.e. $P(\bar{x} > 20:$ hypothesis B is true$)$.

What inference have you reached regarding the source of this second article?

Section 9.5   A More Formalised Procedure for Testing Hypotheses

In the example used so far in this chapter we have had two clear alternatives: the unidentified articles came from either *The Guardian* (hypothesis A) or *Today* (hypothesis B).

Prior to any statistical analysis of the unidentified article, we have no reason to favour either of the two hypotheses. In a formal hypothesis test, it is usual to select one favoured hypothesis (usually representing the theoretical status quo) to be called the *null* hypothesis and usually denoted by H_0. The other hypothesis is defined as the *alternative* hypothesis and is denoted by H_1.

Although there is equal justification for choosing either of the hypotheses A or B for H_0, we shall in fact pick hypothesis A. So we now have

H_0: the article comes from *The Guardian*
H_1: the article comes from *Today*

While we have expressed our hypotheses in words, we now need to find mathematical expressions for H_0 or H_1 in terms of the test statistic that we intend to use to make our decision. In this case, we shall be testing whether $\bar{x}$, the mean number of words per sentence, could have come from either of the sampling distributions for the two newspapers.

H_0: $\bar{x}$ comes from the sampling distribution with $\mu = 23.68$
H_1: $\bar{x}$ comes from the sampling distribution with $\mu = 18.17$

It is quicker to write

H_0: $\mu = 23.68$
H_1: $\mu = 18.17$

In most hypothesis-testing situations, there may not be a precise alternative hypothesis and in fact only the null hypothesis needs to be defined in definite terms. A common testing situation could be

$H_0: \mu = 23.68$
$H_1: \mu \neq 23.68$

Next, we should decide on our testing procedure. As H_0 is the theoretically favoured hypothesis, we shall only reject H_0 if we find statistical results which are at variance with H_0. This will make it relatively difficult for us to reject H_0 and thereby to accept H_1, the alternative hypothesis. However, when analysing our results, we have to take in to account the amount of variability that we can reasonably expect among random samples. We shall only reject H_0 if we find a value of $\bar{x}$ which is unlikely under H_0 but is otherwise in favour of H_1.

We know there is a range of $1.96 \times$ standard error each side of the population mean, within which 95% of all possible sample means will lie (Figure 9.9). (The standard error is the standard deviation of the possible values of the sample mean, i.e. $\sigma/\sqrt{n}$.)

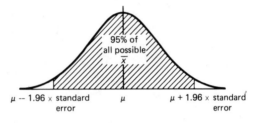

Figure 9.9   Sampling distribution of $\bar{X}$

(Even so, 5% of all possible samples will generate values of $\bar{x}$ which are outside that range.)

We could decide therefore to reject H_0 for values of $\bar{x}$ which fall outside $\mu \pm 1.96 \times$ standard error and not to reject H_0 if $\bar{x}$ lies inside that range (Figure 9.10).

The critical region of size 5% is thus the region for which $|\bar{x}| > \mu + 1.96 \times$ standard error.

The cross-hatched area in Figure 9.10 shows the probability of our obtaining values of $\bar{x}$ which would cause us to reject H_0. The corresponding values of $\bar{x}$ give the *critical region*. The probability that $\bar{x}$ is in the critical region (if H_0 is true) is 5%; so we are working at the 5% level of significance.

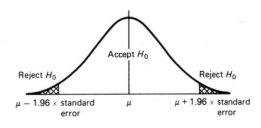

Figure 9.10   Range of rejection of H_0

We have made it quite difficult to reject H_0 but even so, if a large number of samples were tested using these criteria, we could expect to be in error in 5% of cases, and we would reject H_0 when it is in fact true. This is called a type 1 error. If we wish to reduce the probability of making a type 1 error to 0.01, say, we would have to reduce our critical region to $\mu \pm 2.58 \times$ standard error.

Newspaper Articles Example

Criteria for Accepting H_0

$H_0: \mu = 23.68$
$H_1: \mu = 18.17$

The distribution of $\bar{X}$ under H_0 (for samples of size $n = 19$) needs to be considered.

As our alternative hypothesis states a value of μ which is lower than 23.68, it is more sensible to adopt testing criteria which will place all the 5% critical region at the lower end of the distribution under H_0. For extremely low values of $\bar{x}$, it will be more likely that H_1 is true (but for extremely high values we do not wish to favour H_1 over H_0). This is called a one-tailed test (Figure 9.11) (as opposed to the two-tailed test outlined earlier).

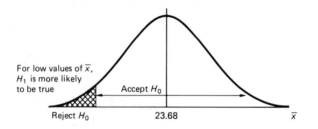

For low values of $\bar{x}$, H_1 is more likely to be true

Accept H_0

Reject H_0

23.68

$\bar{x}$

Figure 9.11   Under H_0: $\bar{X} \sim N(23.68, 8.73)$: ▨, critical region (5%)

The critical region now includes values of $\bar{x}$ which are $\mu - 1.645 \times$ standard error or less. (The value of 1.645 comes from standard normal tables.)

We shall accept H_0 only if $\bar{x}$ is greater than

$\mu - 1.645 \times$ standard error

$$= \mu - 1.645 \times 2.95$$

$$= \mu - 4.85$$

$$= 23.68 - 4.85$$

$$= 18.83$$

If $\bar{x}$ is less than 18.83, we shall reject H_0. (The probability of a type 1 error is 0.05.)

While we have carefully considered the possible error of rejecting H_0 when it is in fact true, there is another kind of error involved in this type of testing. This is the error of accepting the null hypothesis H_0 when the alternative hypothesis H_1 is true. This is called a type 2 error. In order to calculate the probability that this error occurs, we shall need to be able to draw the distribution of $\bar{X}$ according to H_1.

Under H_1, $\bar{X} \sim N(18.17, 4.16)$ and we have already decided to accept H_0 if $\bar{x} > 18.83$ (Figure 9.12).

The probability that we require is represented by the area under the curve, to the right of 18.83. To convert this to a standardised normal variable, we must express 18.83 in terms of the standard error:

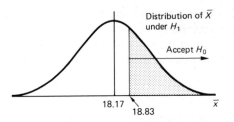

Figure 9.12   Under H_1: $\bar{X} \sim N(18.17, 4.16)$: ▨, probability of a type 2 error

$$z = \frac{18.83 - 18.17}{2.04}$$

$$= \frac{0.66}{2.04}$$

$$= 0.3235$$

Therefore,

area to the left of $z = 0.3235 = 0.626$

and

area to the right $= 0.374$

The probability of a type 2 error is over 37%, while the probability of a type 1 error is only 5%. This means that, if we take 18.83 as our cut-off point for making a decision, the probability of concluding that the article comes from *Today* when it really comes from *The Guardian* (a type 1 error) is fixed at 5% but the probability of concluding that it comes from *The Guardian* when it actually comes from *Today* (a type 2 error) is 37%. In this situation, you may feel that it may be a good idea to try to make the probabilities of these two types of error more equal (Figure 9.13).

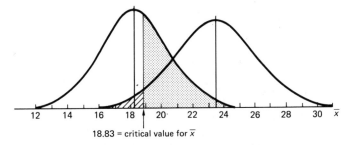

Figure 9.13   Probabilities of errors: ▨, probability of a type 1 error; ▨, probability of a type 2 error

In order to decrease the probability of a type 2 error, we must move our cut-off point (or critical value of $\bar{x}$) to a value higher than 18.83. This will have the effect of increasing the probability of a type 1 error.

Type 1 and Type 2 Errors

The probability of a type 1 error (rejecting H_0 when it is true) is decided by the researcher in advance of any statistical testing. In many situations, however, the probability of a type 2 error (accepting H_0 when it is false) cannot be calculated at all. This is because the distribution of the test statistic ($\bar{X}$ in our example) under H_1 is often not known. We may not have any clear idea of a definite alternative hypothesis. In these cases we have hypotheses such as

H_0: $\bar{x}$ comes from the sampling distribution with $\mu = 23.68$
H_1: $\bar{x}$ does *not* come from the sampling distribution with $\mu = 23.68$

You may find the Table 9.5 useful when trying to remember the nature of type 1 and type 2 errors.

Table 9.5   Nature of type 1 and type 2 errors

		Actual situation	
		H_0 is true	H_1 is true
Conclusion from test	Accept H_0	No error	Type 2 error
	Reject H_0	Type 1 error	No error

Section 9.6   (Some) Solutions

Section 9.4

The mean number of words per sentence is $\bar{x} = 22.4$.

The variance of the sampling distribution for *The Guardian* is

$$\frac{\sigma^2}{n} = \frac{165.89}{17} = 9.758$$

with a standard error of 3.12; the variance of the sampling distribution for *Today* is

$$\frac{\sigma^2}{n} = \frac{79.03}{17} = 5.708$$

with a standard error of 2.39.

Acknowledgements

'How fall-out clouds the political climate' appeared as 'Alternative View' in *Today* and was written by Ben Pimlott in 1986.

'P.C. suspended after gun "horseplay"' appeared in 1986 in *The Guardian* and was written by Paul Hoyland.

Chapter 10 Hypothesis Testing II

Section 10.1 Test of a Single Mean (Large Sample)

Practical 10.1 Packets of Crisps

Problem: Do packets of crisps contain the correct weight?

Tesco supermarket 'own brand' plain crisps are sold in small packets which state 25 g e. The e means that the distribution of weights should form a normal distribution with a mean mass of 25 g. (Therefore, we could find that a sample of packets of crisps could include roughly half which weigh under 25 g.) As previously we shall use the term weight colloquially in the text when technically we mean mass. We weigh the crisps to find their mass in grams.

Equipment and Materials

Buy a number of packets of crisps all one brand and flavour (aim to get at least 30 packets). You will need an electronic balance (accurate to 0.01 g); so the weighing will probably need to be done in a laboratory.

Method

Weigh each full packet of crisps; then tip the crisps out into a spare container, and weigh the empty packet. The contents may be calculated by subtracting the two weights. This method is easier and less messy than weighing the crisps themselves.

The Null Hypothesis

The mean mass of crisps contained in Tesco packets is 25 g.

$H_0: \mu = 25$ g
$H_1: \mu \neq 25$ g

This is a two-tailed test.

The Test Statistic

In order to test the null hypothesis, you need to weigh the contents of a number of packets of crisps and to calculate the mean quantity.

Level of Significance

The level of significance is an arbitrary choice to a certain extent. If you work with a 5% level of significance the probability of making a type 1 error (of rejecting H_0 when it is true) will be 0.05. The probability of a type 2 error—accepting the null hypothesis when the alternative hypothesis is true—cannot be calculated as we do not know a precise value for μ under H_1.

Testing at the 5% Level

We shall reject H_0 if we have a sample mean lying outside the range $\mu \pm 1.96 \times$ standard error.

Distribution of $\bar{X}$ under H_0

The distribution of $\bar{X}$ under H_0 is given in Figure 10.1.

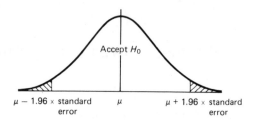

Figure 10.1 Distribution of $\bar{x}$ under H_0: ▨, critical region (reject H_0).

The standard error of $\bar{X} = \sigma/\sqrt{n}$ or $\sqrt{\sigma^2/n}$.

Our Results

We weighed 48 packets of Tesco ready-salted crisps. The mean mass of the contents was 25.28 g and the variance only 0.0599 g (the raw data are reproduced at the end of this section in Table 10.1), i.e.

$\bar{x} = 25.28$ g

$s^2 = 0.0599$ g

(variance with divisor $n - 1$).

The Significance Test

H_0: $\mu = 25$ g
H_1: $\mu \neq 25$ g

For a two-tailed test at the 5% level of significance, the critical value for Z is $|Z| > 1.96$.

$$z = \frac{\bar{x} - \mu}{\sqrt{\sigma^2/n}}$$

$$z = \frac{25.28 - 25}{\sqrt{0.0599/48}}$$

$$= \frac{0.28}{0.035}$$

$$= 8$$

This value of z is highly significant and we must reject H_0 under the testing criteria that we have set up. The extreme significance of the result is caused by the fact that the variance of the contents is so small. The machinery used to fill the packets must be calibrated very accurately. In practical terms, any consumer 'watch dog' organisation will not be worried by any measured difference which exceeds the stated weight. Also the manufacturers prefer to maintain good customer relations and, while the statistical result suggests that their packets are significantly overweight, in practical terms an extra 0.28 g on average is only 1.12% of the stated packet weight. Statistical significance does not always imply practical significance!

Section 10.2 Alternative Practicals

Practical 10.2 Weight of Contents of Pre-packed Food

The experiment can be adapted to test the weight of the contents of any pre-packed food.

Table 10.1 Masses of 48 packets of crisps with a normal mass of 25 g

Packet number	Full packet (g)	Empty packet (g)	Crisps only (g)
1	26.28	1.45	24.83
2	27.00	1.28	25.72
3	26.55	1.27	25.28
4	26.44	1.25	25.19
5	26.70	1.35	25.35
6	26.26	1.25	25.01
7	26.31	1.31	25.00
8	26.44	1.21	25.23
9	26.17	1.32	24.85
10	26.32	1.29	25.03
11	26.66	1.38	25.28
12	26.28	1.24	25.04
13	26.86	1.25	25.61
14	26.45	1.21	25.24
15	26.44	1.27	25.17
16	26.28	1.20	25.08
17	26.58	1.21	25.37
18	26.69	1.35	25.34
19	26.53	1.19	25.34
20	26.81	1.28	25.53
21	26.26	1.24	25.02
22	26.79	1.25	25.54
23	26.75	1.33	25.42
24	26.43	1.24	25.19
25	27.00	1.22	25.78
26	26.68	1.19	25.49
27	26.37	1.26	25.11
28	26.53	1.24	25.29
29	26.48	1.37	25.11
30	26.32	1.24	25.08
31	26.28	1.26	25.02
32	26.97	1.30	25.67
33	26.91	1.28	25.63
34	26.51	1.24	25.27
35	27.03	1.25	25.78
36	26.37	1.23	25.14
37	26.98	1.34	25.64
38	26.75	1.22	25.53
39	26.10	1.19	24.91
40	26.59	1.28	25.31
41	26.57	1.26	25.31
42	26.91	1.23	25.68
43	26.41	1.31	25.10
44	26.45	1.30	25.15
45	26.53	1.31	25.22
46	26.77	1.25	25.52
47	26.38	1.17	25.21
48	26.59	1.30	25.29

Practical **10.3   Contents of Matchboxes**

Count the matches in each of a large sample of matchboxes to see whether the average contents varies significantly from that stated.

Practical **10.4   Heights of Girls or Boys**

Measure the heights of a sample of girls or boys aged 16–19 years. The 1980 OPCS survey (referred to in Chapter 5) found the following results:

mean height for females aged 16–19 years = 161.7 cm
mean height for males aged 16–19 years = 174.6 cm

Test whether the mean of your sample is significantly different.

Section **10.3**   **Significance Test Based on a Binomial Model (Small Sample)**

Practical **10.5   Telepathy**

Problem: **Are you telepathic?**

Equipment

You require the following.

2 persons for each trial.
1 overseer.
3 chairs.
1 desk.
1 screen.
1 stopwatch.
1 pencil and paper.
1 pack of ordinary playing cards.
A table of random numbers, or a calculator which generates random numbers.

Figure 10.2   Arrangement of personnel for Practical 10.5

Method

There are several versions of this experiment, but this is one of the simplest and requires no special equipment.

Arrange your personnel as in Figure 10.2.

Prior to the experiment the overseer uses random numbers to select 12 cards and to determine in which order they will be presented during the experiment. The suit is the most important characteristic to be identified, and the denominations used do not matter. The random numbers may be used in any scheme similar to the following:

0, 1	club
2, 3	diamond
4, 5	heart
6, 7	spade
8, 9	these numbers should be ignored

The overseer then passes the pile of cards (face down) to subject 1 and then announces 'First card'. Subject 1 turns the first card over and concentrates on it for about 30 s.

Subject 2 must decide on the *suit* of that card and write it down. When this has been done, the overseer should announce 'Second card' and the experiment continues as before.

If there is an audience present, they must not communicate with any of the participants. At the end, count up the number of correct responses by subject 2.

The Null Hypothesis

The null hypothesis is that telepathy does not exist and that subject 2 cannot determine the suit of the card by reading the thoughts of subject 1 and can only guess at the answer. If he or she is guessing the answer, the probability of being correct is 1/4. If they are telepathic, the probability of being correct will be more than 1/4. Therefore,

$$H_0: p = \frac{1}{4}$$

$$H_1: p > \frac{1}{4}$$

This is a one-tailed test. (We are assuming that anti-telepathy cannot exist, and we shall wish to reject H_0 only if the subject has a high number of correct answers.)

The Test Statistic

The test statistic X is the number of correct answers given by subject 2. Under H_0, X has a distribution $B(12, 1/4)$ as 12 cards have been presented and there is a probability of 1/4 of a correct answer each time. We can calculate the probability associated with each value of X using the binomial model (Table 10.2 and Figure 10.3).

If subject 2 gets a high number of correct answers, we shall wish to reject H_0 and opt for H_1 as being more likely. To work with a 5% level of significance, we must determine the extent of the critical region so that it contains a total probability of 0.05 at the most. The values of X in the critical region will be high values at the top end of the distribution.

The Significance Test

Since

$$P(X \geq 7) = 0.014\,34$$

$$P(X \geq 6) = 0.054\,44$$

Table 10.2   Binomial probabilities ($n = 12$; $p = 1/4$; $q = 3/4$)

Number x of correct responses	$p(x)$	
0	$\left(\dfrac{3}{4}\right)^{12}$	≈ 0.0317
1	$12 \times \left(\dfrac{3}{4}\right)^{11} \times \left(\dfrac{1}{4}\right)$	≈ 0.1267
2	$\dfrac{12 \times 11}{2 \times 1} \times \left(\dfrac{3}{4}\right)^{10} \times \left(\dfrac{1}{4}\right)^{2}$	≈ 0.2323
3	$\dfrac{12 \times 11 \times 10}{3 \times 2 \times 1} \times \left(\dfrac{3}{4}\right)^{9} \times \left(\dfrac{1}{4}\right)^{3}$	≈ 0.2581
4	$\dfrac{12 \times 11 \times 10 \times 9}{4 \times 3 \times 2 \times 1} \times \left(\dfrac{3}{4}\right)^{8} \times \left(\dfrac{1}{4}\right)^{4}$	≈ 0.1936
5	$\dfrac{12 \times 11 \times 10 \times 9 \times 8}{5 \times 4 \times 3 \times 2 \times 1} \times \left(\dfrac{3}{4}\right)^{7} \times \left(\dfrac{1}{4}\right)^{5}$	≈ 0.1032
6	$\dfrac{12 \times 11 \times 10 \times 9 \times 8 \times 7}{6 \times 5 \times 4 \times 3 \times 2 \times 1} \times \left(\dfrac{3}{4}\right)^{6} \times \left(\dfrac{1}{4}\right)^{6}$	≈ 0.0401
7	$\dfrac{12 \times 11 \times 10 \times 9 \times 8}{5 \times 4 \times 3 \times 2 \times 1} \times \left(\dfrac{3}{4}\right)^{5} \times \left(\dfrac{1}{4}\right)^{7}$	≈ 0.0115
8	$\dfrac{12 \times 11 \times 10 \times 9}{4 \times 3 \times 2 \times 1} \times \left(\dfrac{3}{4}\right)^{4} \times \left(\dfrac{1}{4}\right)^{8}$	≈ 0.0024
9	$\dfrac{12 \times 11 \times 10}{3 \times 2 \times 1} \times \left(\dfrac{3}{4}\right)^{3} \times \left(\dfrac{1}{4}\right)^{9}$	≈ 0.0004
10	$\dfrac{12 \times 11}{2 \times 1} \times \left(\dfrac{3}{4}\right)^{2} \times \left(\dfrac{1}{4}\right)^{10}$	$\approx 0.000\,04$
11	$12 \times \left(\dfrac{3}{4}\right) \times \left(\dfrac{1}{4}\right)^{11}$	$\approx 0.000\,002$
12	$\left(\dfrac{1}{4}\right)^{12}$	≈ 0

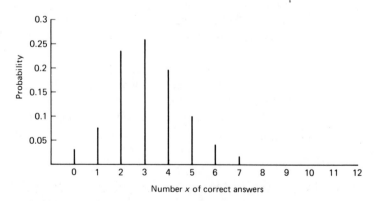

Figure 10.3   Probability distribution of X under H_0

the probability of 7 or more correct answers is 0.014 (1% approximately) which is well below 5%. Strictly speaking, we should reject H_0 for that range if we want the probability of a type 1 error to be at most 5%, but this is giving too much credence to a magical '5%'. If we are prepared to go slightly over 5%, the critical region can be extended to include 6 or more correct answers.

So, if subject 2 has less than 6 correct answers, you will retain the null hypothesis and conclude that any correct answers were obtained by guessing. For a higher number of correct answers, you might be prepared to concede that it is unlikely that your subject is guessing. Are the two people colluding or cheating in some way? Or are they telepathic?

Section 10.4   An Alternative Practical

Practical 10.6   The Triangle Taste Test

Problem: Can people distinguish between margarine and butter (or Coke and Pepsi Cola, etc.)?

Method

Each subject is presented with three samples, two from one product and one from the other, and is asked to pick out the odd sample. These samples should be as alike as possible, i.e. the sizes, shapes and temperatures should be the same. If there is a colour difference, try to minimise it with coloured lighting or containers.

It has been found that people generally have a tendency to pick out the middle one of three samples (unless there is a marked difference between them). Also the first sample seems to create a stronger impression than those which follow. These are sources of bias which we can try to eliminate in the following ways.

1. Present the odd sample in the left-hand position to one-third of the subjects, and in the middle and right-hand positions to one-third. No subject should know which type of sample has been given. Make up one-third of each type and use random numbers to allocate them to the subjects.

   Keep a careful record of each trial. Toss a coin to decide which product is to be presented twice.
2. Each subject should take a small sip of water before starting the experiment and between each tasting.

   Do not allow any collaboration between subjects and carry out the tests in quiet surroundings free from odours.

   Each subject should be asked the following two questions (tell them that one sample will be different).

   a. Can you detect any difference in the flavours?
   b. Which sample do you think is different?

Analysis

Ignore all subjects who could not detect any difference. Count the number n who could detect a difference and count the number x who correctly identified the odd sample.

The Null Hypothesis

The null hypothesis is that there is no discernible difference between the subjects and that subjects can only identify the odd sample by guessing:

$$H_0: p = \frac{1}{3}$$

$$H_1: p > \frac{1}{3}$$

This is a one-tailed test.

If a small number of subjects is used, the binomial distribution may be used as a model to determine whether or not to accept H_0 on the basis of your trials. The probabilities for each value of X may be calculated as terms in the distribution $B(n, 1/3)$. See Section 10.3 for the method of analysis.

For a larger number of subjects, use the normal approximation to the binomial distribution.

Section 10.5   Comparison of Two Sample Means (Large Samples)

Practical 10.7   Estimation of Time

Problem: Do females and males estimate time intervals differently?

Method

Select two samples, one of males and one of females, each with similar age distributions. Preferably, these should be random samples with about 50 subjects in each, although you do not require equal numbers of males and females.

Ask each person individually to estimate 1 min. Making sure not to give them any help or visual clues, time their estimate with a stopwatch, digital watch or watch with a second hand. Record the estimates given by females and males separately.

The Null Hypothesis

The times estimated by females and males are not significantly different. We assume that females' estimates of 1 min have a distribution with mean μ_1 and variance σ_1^2 and that males' estimates have a distribution with mean μ_2 and variance σ_2^2. Therefore,

$$H_0: \mu_1 = \mu_2$$

$$H_1: \mu_1 \neq \mu_2$$

We shall test the difference between the two means; so the null hypothesis may be rewritten as

$$H_0: \mu_1 - \mu_2 = 0$$

$$H_1: \mu_1 - \mu_2 \neq 0$$

This is a two-tailed test.

The Test Statistic

Under H_0, both sample means $\bar{x}_1$ and $\bar{x}_2$ come from sampling distributions with mean μ and variance σ_1^2/n_1 and σ_2^2/n_2. If we assume also that the samples are large enough we have, by the central limit theorem, the distributions in Figure 10.4.

To determine the distribution of the difference in means, use the result

$$E(\bar{X}_1 - \bar{X}_2) = E(\bar{X}_1) - E(\bar{X}_2)$$

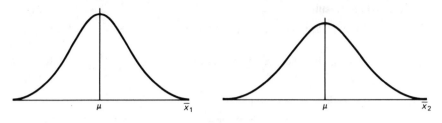

Figure 10.4 $\bar{X}_1 \sim N(\mu, \sigma_1^2/n_1)$ and $\bar{X}_2 \sim N(\mu, \sigma_2^2/n_2)$

giving, under the null hypothesis,

$$E(\bar{X}_1 - \bar{X}_2) = \mu - \mu = 0$$

We also have

$$\text{Var}(\bar{X}_1 - \bar{X}_2) = \text{Var}(\bar{X}_1) + \text{Var}(\bar{X}_2)$$

Since $\bar{X}_1$ and $\bar{X}_2$ are independent,

$$\text{Var}(\bar{X}_1 - \bar{X}_2) = \frac{\sigma_1^2}{n_1} + \frac{\sigma_2^2}{n_2}$$

Thus under H_0 the distribution of $\bar{X}_1 - \bar{X}_2$ is $N(0, \sigma_1^2/n_1 + \sigma_2^2/n_2)$.

Note

The solution given here does *not* assume that the samples were originally taken from the same population so that X_1 and X_2 have the same variance. For a solution of that type, see Section 10.6.

We shall accept the alternative hypothesis if we get high or low values of $\bar{x}_1 - \bar{x}_2$. The critical region is at the extreme ends of the distribution. If we work at a 5% level of significance, the critical value for Z is $|Z| > 1.96$ (Figure 10.5).

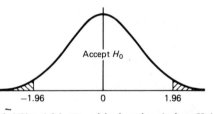

Figure 10.5 Distribution with $|Z| > 1.96$: ▨, critical region (reject H_0)

$Z \sim N(0, 1)$ where z is calculated as

$$z = \frac{\bar{x}_1 - \bar{x}_2}{\sqrt{\sigma_1^2/n_1 + \sigma_2^2/n_2}}$$

In practice, we do not usually know σ_1^2 and σ_2^2 and so use s_1^2 and s_2^2 calculated from the sample. Provided that the sample sizes are large the distribution of

$$\frac{\bar{X}_1 - \bar{X}_2}{\sqrt{s_1^2/n_1 + s_2^2/n_2}}$$

will still be near enough to normal.

Our Results

We took random samples of 85 females and 53 males. Our results are given in Tables 10.3 and 10.4.

Note that we use s^2, the sample estimate of σ^2, for each sample.

Table 10.3 Times estimated

True time estimated by subject (s)	Frequency	
	Females	Males
20–24	2	1
25–29	2	3
30–34	8	5
35–39	8	9
40–44	9	10
45–49	13	9
50–54	16	7
55–59	10	5
60–64	9	2
65–69	4	1
70–74	3	1
75–79	1	

Table 10.4 Variance

Group	n	$\bar{x}$	Variance s^2
Females	85	48.94	147.10
Males	53	51.72	120.11

The Significance Test

$H_0: \mu_1 - \mu_2 = 0$
$H_1: \mu_1 - \mu_2 \neq 0$

Using a two-tailed test at the 5% level of significance the critical value for Z is $|Z| > 1.96$.

$$z = \frac{\bar{x}_1 - \bar{x}_2}{\sqrt{\sigma_1^2/n_1 + \sigma_2^2/n_2}}$$

$$= \frac{48.94 - 51.72}{\sqrt{147.10/85 + 120.11/53}}$$

$$\approx -\frac{2.78}{2.00}$$

$$= -1.39$$

This value of z is not significant at the 5% level and so we must conclude that at this level there is no difference between the two means, i.e. females and males have not made significantly different estimates for 1 min.

Section **10.6   Alternative Solution: Comparison of Two Means (Large Samples; assuming that they come from populations with the same variance)**

This is an alternative solution to the problem outlined in Section 10.5. The experiment remains the same, but we assume that females' estimates have a distribution with mean μ_1 and variance σ^2; the males' estimates have a distribution with mean μ_2 and variance σ^2.

The Null Hypothesis

$H_0: \mu_1 = \mu_2$, i.e. $\mu_1 - \mu_2 = 0$

$H_1: \mu_1 - \mu_2 \neq 0$

If the samples come from populations with the same means and variances, and the samples are large then, by the central limit theorem,

$$\bar{X}_1 - \bar{X}_2 \sim N\left(0, \frac{\sigma^2}{n_1} + \frac{\sigma^2}{n_2}\right)$$

As σ^2 is unknown, the two sample variances are combined to give an estimate of the population variance. This is denoted by s^2 where

$$s^2 = \frac{n_1 \, \mathrm{Var}_1 + n_2 \, \mathrm{Var}_2}{n_1 + n_2 - 2} \qquad \text{or} \qquad \frac{(n_1 - 1){s_1}^2 + (n_2 - 1){s_2}^2}{n_1 + n_2 - 2}$$

Using the results of our study given in Section 10.5, we get

$$s^2 = \frac{84 \times 147.10 + 52 \times 120.11}{85 + 53 - 2}$$

$$= \frac{12\,356.4 + 6245.72}{136}$$

$$= \frac{18\,602.12}{136}$$

Therefore,

estimate of population variance $s^2 = 136.78$

Using this value, we can estimate the variance for the distribution of the differences in sample means:

$$\mathrm{Var}(\bar{X}_1 - \bar{X}_2) = \frac{s^2}{n_1} + \frac{s^2}{n_2}$$

$$= \frac{136.78}{85} + \frac{136.78}{53}$$

$$\approx 1.6092 + 1.5808$$

$$= 4.1900$$

Note

If you wish to test whether ${\sigma_1}^2 = {\sigma_2}^2$ before proceeding to combine them for an estimate of σ^2, you will need to use the F ratio test (see Section 10.11).

The Significance Test

$H_0: \mu_1 - \mu_2 = 0$ and $\sigma_1{}^2 = \sigma_2{}^2 = \sigma^2$

$H_1: \mu_1 - \mu_2 \neq 0$

We take as our test statistic

$$z = \frac{\bar{x}_1 - \bar{x}_2}{\sqrt{s^2/n_1 + s^2/n_2}}$$

If we had not had to use s^2 to estimate σ^2, Z would have an $N(0, 1)$ distribution. The estimating of σ^2 by s^2 means that we strictly speaking have a different distribution but fortunately, for large samples, this is almost the same as the normal. (For small samples see Section 10.9.)

Using a two-tailed test at the 5% level of significance the critical value for Z is $|Z| > 1.96$.

$$z = \frac{48.94 - 51.72}{\sqrt{4.190}}$$

$$\approx \frac{-2.78}{2.047}$$

$$= 1.36 \text{ (to two decimal places)}$$

This value of z is not significant and we retain H_0. The females and males in our samples have not made significantly different estimates of 1 minute.

Section 10.7   Alternative Practicals

Practical 10.8   **Reactions I**

Problem: **Do sixth-formers and middle-aged people react equally quickly?**

Equipment

To conduct the experiment you will need a reaction ruler developed by the Centre for Statistical Education and available from many suppliers of school equipment. Alternatively, you may use an ordinary 30 cm ruler, and Table 10.5 can be used to convert distance in millimetres to reaction time in hundredths of a second.

Table 10.5   Conversion of distances to reaction times

Reaction time (10^{-2} s)	5	6	7	8	9	10	11	12	13	14	15	16	17	18	19	20	21	22	23	24
Distance (mm)	12.3	17.6	24	31.4	40	49	59	71	83	96	110	125	142	159	177	196	216	237	259	282

Method

One pupil holds the ruler suspended vertically by thumb and forefinger. The ruler should be held on the 30 cm line. A second pupil is ready to catch the ruler with his thumb and forefinger over the 0 cm line but not touching the ruler. The first pupil chooses when to drop the ruler at any time up to 20 s from the start of the experiment. As the first pupil drops the ruler, the second tries to catch it. The distance that the ruler has fallen is recorded and from this the reaction time may be calculated or read directly from the ruler.

Conduct this experiment for two large samples of sixth-formers and middle-aged people. Since people improve with practice, you must either take each person's first result or give everyone an equal opportunity for practice.

Practical 10.9   Mental Abilities of Males and Females

Problem: **Do males and females possess differing mental abilities?**

It is believed that men are better at visual tasks and women are better at verbal tasks, and this experiment is designed to test these hypotheses.

No special equipment is required apart from a stopwatch. First ask your subject to count the number of letters in the alphabet which contain the sound 'ee'. This is the verbal task. Record the time taken in seconds to complete the task (and also the number of mistakes, which can be analysed separately).

The second task is the visual task, and for this ask your subject to count the number of letters in the alphabet which contain a curved line in their upper case (capital letter) form. Again record the time taken (and the number of errors).

After calculating the mean time taken for each task separately by the males and the females, various hypotheses may be tested.

1. Do females complete the verbal task more quickly than the visual task?
2. Do males complete the visual task more quickly than the verbal task?
3. Do females complete the verbal task more quickly than males do?
4. Do males complete the visual task more quickly than females do?

You will need large samples of males and females in order to use the method of analysis outlined in Section 10.5 or 10.6. If you can only obtain results for small samples you must use the *t* distribution to test for a difference in sample means as shown in Section 10.9.

Practical 10.10   Lengths of Sentences in Newspapers

Problem: **Does *The Guardian* use longer sentences than *Today*?**

Take random samples of about 150 sentences from both newspapers, and find the mean and variance of the number of words per sentence. You will have sufficient data to test this hypothesis if you have worked through Chapter 9.

You may adapt this study to compare the lengths of words or sentences used in any two publications or books.

Section 10.8   Test of Two Sample Proportions (Large Samples)

Practical 10.11   Conservation of Volume

Problem: Does a larger proportion of older pupils than younger pupils understand conservation of volume?

Conservation is an important idea in mathematical thinking. Young children cannot make any progress with simple arithmetical problems until they understand the concept of conservation of number. That is, the constancy of a number or quantity of objects, which retains the same value unless something is added or taken away. (Five counters in a row are the same as five counters arranged in any other pattern.) As mathematical experience and thinking develop, the child comes to understand conservation of area and conservation of volume.

Equipment

You require two large balls of Plasticine of the same weight, a simple see-saw balance, two identical glass beakers containing sufficient water for a ball of Plasticine to be immersed without any spillage, and the written instructions.

We suggest that you conduct the experiment once with a class of young children (first-year secondary or younger) and then again with older pupils (fourth-year secondary or older).

Read the instructions aloud to the class and ask the pupils to write down their answers, either on plain paper or on prepared answer sheets. The experimenter should not give the pupils any clues nor should they be allowed to discuss the problems or to help each other.

The pupils' questions (which you may wish to print on a sheet for each child) are in ordinary type, while your instructions and additional comments are in italic type.

Stand at the front of the class with the balance on the teacher's desk. All the pupils must be able to see clearly what you do. Give them adequate time to write down their answers.

Verbal Instructions for the Experiment

Prior to the experiment, say
This is not a test with right or wrong answers to any of the questions. We are only interested in the reasons you give for your answers. There is no need to worry about the questions nor to copy answers from anyone else. If you do not know the answer to a question, write 'Don't know'.

Show the pupils the two balls of Plasticine.
Here are two balls of Plasticine which I am going to place on the balance.
They should balance—if the weights are not equal remove small quantities from one ball until equilibrium is achieved.

1. Do the two balls of Plasticine weigh the same?

Take the balls of Plasticine off the balance.

2. Do the balls of Plasticine have the same volume?
3. I have made one of the balls into a sausage shape. Will the two pieces, the ball and the sausage, weigh the same?

Roll one ball of Plasticine into a sausage shape and leave the other as it is. Do not replace them on the balance. You may repeat the question.

4. Explain your answer.
5. Do the ball and the sausage shape have the same volume?
6. Explain your answer.
7. Suppose that I rolled the sausage back into a ball, would the balls of Plasticine have the same volume?

Do not roll the sausage shape back into a ball.

8. Explain your answer.

Place the two identical beakers of water on the desk. There should be plenty of space above the water level so that it is clear that they would not overflow. Say Notice that the water in these two glass beakers reaches the same height.

9. What would happen to the water levels if I placed the ball of Plasticine in the left-hand beaker, and the sausage shape in the right-hand beaker so that both are completely under the water?

Do not put the Plasticine in the water!

10. Explain your answer.

The Null Hypothesis

We assume that the younger pupils are drawn from a population with proportion p_1 who understand the concept of conservation of volume; for the older pupils the proportion is p_2. Our null hypothesis is that there is no difference in the proportion of pupils in the two age groups who have acquired the concept of conservation of volume.

H_0: $p_1 = p_2$ or $p_2 - p_1 = 0$

H_1: $p_2 > p_1$ or $p_2 - p_1 > 0$

As the second group of pupils are older, the alternative hypothesis favours the fact that a higher proportion of these pupils will understand the concept; so we use a one-tailed test.

The Test Statistic

We used the proportion of pupils correctly answering question 5 as our test statistic. You might also consider separately questions 3 and 9, or test for differences between them.

We calculated r_1/n_1, the proportion of younger children correctly answering question 5, and r_2/n_2, the proportion of older children correctly answering question 5.

Under H_0, both samples come from the same population in which the actual proportion is p. Both sample proportions are values from sampling distributions which are normally distributed. Using the normal approximation to binomial,

$$\frac{R_1}{n_1} \sim N\left(p, \frac{p(1-p)}{n_2}\right)$$

$$\frac{R_2}{n_2} \sim N\left(p, \frac{p(1-p)}{n_2}\right)$$

where n_1 and n_2 are the respective sample sizes.

Under H_0,

$$E\left(\frac{R_2}{n_2} - \frac{R_1}{n_1}\right) = E\left(\frac{R_2}{n_2}\right) - \left(\frac{R_1}{n_1}\right)$$

$$= p - p$$

$$= 0$$

$$\mathrm{Var}\left(\frac{R_2}{n_2} - \frac{R_2}{n_1}\right) = \mathrm{Var}\left(\frac{R_2}{n_2}\right) + \mathrm{Var}\left(\frac{R_2}{n_1}\right)$$

$$= \frac{p(1-p)}{n_1} + \frac{p(1-p)}{n_2}$$

Our best estimate of p is obtained by pooling the results from both samples:

$$\text{estimate of } p = \frac{\text{number of children who answered correctly}}{\text{total number of children}}$$

$$= \frac{r_1 + r_2}{n_1 + n_2}$$

The Level of Significance

Testing at the 5% level, with a one-tailed test the critical value for Z will be $Z > 1.645$. Under H_0,

$$\frac{R_2}{n_1} - \frac{R_1}{n_1} \sim N\left(0, \frac{p(1-p)}{n_1} + \frac{p(1-p)}{n_2}\right)$$

The distribution is shown in Figure 10.6.

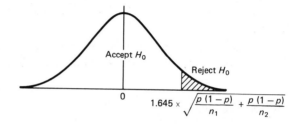

Figure 10.6 $R_2/n_1 - R_1/n_1 \sim N(0, p(1-p)/n_1 + p(1-p)/n_2)$: ▨, critical region (5%)

Our Results

Among a class of 26 11–12 year olds, 17 pupils correctly answered question 5. In a class of 31 14–15 year olds, 25 answered correctly.

$$\frac{r_1}{n_1} = \frac{17}{26} \approx 0.654$$

$$\frac{r_2}{n_2} = \frac{25}{31} \approx 0.806$$

So our pooled two-sample estimate of the population proportion is

$$p = \frac{\text{number of children who answered correctly}}{\text{total number of children}}$$

$$= \frac{17 + 25}{26 + 31}$$

$$= \frac{42}{57}$$

$$\approx 0.737$$

$$1 - p = 0.263$$

The Significance Test

$$H_0: p_2 - p_1 = 0$$
$$H_1: p_2 - p_1 > 0$$

If we had known p, then Z would have an $N(0, 1)$ distribution. However, since we estimated p, this is not the case strictly speaking. We rely on the fact that for large samples the distribution will become nearly normal and so still use the tables of the normal distribution.

For a one-tailed test the critical value for Z at the 5% level is 1.645.

$$z = \frac{r_2/n_2 - r_1/n_1}{\sqrt{p(1-p)/n_1 + p(1-p)/n_2}}$$

$$= \frac{0.806 - 0.654}{\sqrt{(0.737 \times 0.263)/26 + (0.737 \times 0.263)/31}}$$

$$\approx \frac{0.152}{\sqrt{0.00745 + 0.00625}}$$

$$= \frac{0.152}{\sqrt{0.01370}}$$

$$\approx \frac{0.152}{0.1170}$$

$$\approx 1.299$$

This value of z does not lie in the critical region and our result is not significant at the 5% level. On the basis of these samples, we must retain the null hypothesis and conclude that we do not have sufficient statistical evidence to suggest that the older pupils had any better understanding of the conservation of volume than the younger ones.

Practical 10.12   Left-handed Pupils in Top Maths Sets (Alternative Practical)

Is the proportion of left-handed people higher in top maths sets? You can only test this hypothesis if pupils are grouped according to ability for their mathematics lessons.

Section 10.9   Test for the Difference of Two Sample Means (Small Samples)

Practical 10.13   Coaction in Sports

Problem: Does the presence of other participants improve task performance?

While the presence of spectators often improves performance, particularly in competitive sports and games, it has also been suggested that the presence of other participants may have the same effect. For instance it has been observed that bicycle riders cycle faster in groups than they do alone. (This phenomenon is known as coaction.) If this hypothesis is true, it has important implications not only for sport but also in other fields such as education (and even, it could be argued, for the formation of such groups as slimming clubs!).

Method

Divide the class into two groups of equal numbers and sex distribution. Aim for about 10 in each half. Toss a coin to decide which half will perform the task alone, and which half will perform the task in small groups (preferably with equal numbers—3, 4 or 5 students in each group).

The subjects are required to sit on the edge of a bench or chair and extend one leg in a horizontal position for as long as possible. Ensure that they do sit on the very edge of the bench; otherwise the bench or chair will provide support for the leg and the times that you record will be unrealistically long. The time for which the leg is kept extended should be recorded in seconds. Subjects selected to perform the task alone should only have the time recorder present, and that person should not converse with the subject or offer any encouragement.

The Null Hypothesis

We assume that the times taken by subjects on their own follows a normal distribution $N(\mu_1, \sigma^2)$ and that those in groups follow a normal distribution $N(\mu_2, \sigma^2)$. (Note that we are assuming equal variances.) Our null hypothesis is that the performance of subjects working alone is the same as that of subjects working in groups:

$$H_0: \mu_1 = \mu_2 \qquad \text{or} \qquad \mu_2 - \mu_1 = 0$$

Under the alternative hypothesis, the students working in groups should have longer times:

$$H_1: \mu_2 > \mu_1 \qquad \text{or} \qquad \mu_2 - \mu_1 > 0$$

This will require a one-tailed test.

The Test Statistic

The mean time for each of the two samples is calculated.

Because we have assumed the times to be normally distributed, we can test for a difference in the two means using the Student t distribution (W. S. Gossett). The distribution curve for t is different for each value of v, the number of degrees of freedom.

As we are comparing the means of two samples, $v = n_1 + n_2 - 2$. Under H_0,

$$E(\bar{X}_2 - \bar{X}_1) = 0$$

and

$$\text{Var}(\bar{X}_2 - \bar{X}_1) = \frac{\sigma^2}{n_1} + \frac{\sigma^2}{n_2}$$

If the evidence is that the underlying variances are *not* equal, then the following analysis cannot be done for small samples.

You may test whether the variances are equal, using the F ratio test (Section 10.11). If the ratio does not reach a significant value, the two sample variances may be combined to give a pooled estimate of the population variance, where

$$s^2 = \frac{n_1 \, \text{Var}_1 + n_2 \, \text{Var}_2}{n_1 + n_2 - 2} \qquad \text{or} \qquad \frac{(n_1 - 1)s_1^2 + (n_2 - 1)s_2^2}{n_1 + n_2 - 2}$$

and we test the value of

$$t_{n_1 + n_2 - 2} = \frac{\bar{x}_2 - \bar{x}_1}{\sqrt{s^2/n_1 + s^2/n_2}}$$

on a t distribution.

Our Results

Our results are given in Table 10.6.

From these we obtained the statistics in Table 10.7.

Table 10.6   Our results

Alone		Group	
Name	Time (s)	Name	Time (s)
Eddie	97	Andy	400
Pat	203	Sean	400
Kathy	78	Ian	289
Dave	128	Jean	170
Rob	180	Jan	238
Tim	150	Sue	220
Jane	102	John	186
Phil	91	Paul	302

Table 10.7   Means and variances

	Mean	Variance $s^2 = \dfrac{\Sigma(x - \bar{x})^2}{n - 1}$
Alone	128.625	2045.1
Group	275.625	7930.3

Our pooled two-sample estimate of variance is

$$s^2 = \frac{(7 \times 2045.1) + (7 \times 7930.3)}{8 + 8 - 2}$$

$$= \frac{14\,315.7 + 55\,512.1}{14}$$

$$= \frac{69\,827.8}{14}$$

$$= 4987.7$$

The Significance Test

$H_0: \mu_2 - \mu_1 = 0$

$H_1: \mu_2 - \mu_1 > 0$

Using a one-tailed test at the 5% level of significance, the critical value for $t(v = 14)$ is 1.76.

$$t_{n_1 + n_2 - 2} = \frac{\bar{x}_1 - \bar{x}_2}{\sqrt{s^2/n_1 + s^2/n_2}}$$

$$t_{14} = \frac{275.625 - 128.625}{\sqrt{4987.7/8 + 4987.7/8}}$$

$$\approx \frac{147}{\sqrt{623.46 + 623.46}}$$

$$\approx \frac{47}{\sqrt{1246.9}}$$

$$\approx \frac{147}{35.3}$$

$$\approx 4.16$$

This value of t is significant at the 5% level; so we can reject H_0. On the basis of these small samples, we have found some support for the theory that the presence of other participants does improve performance.

Practical 10.14 **Coaction When a Difficult Task is Being Performed (Alternative Practical)**

The task used in Practical 10.13 is very simple, requiring no special skill. More recently, it has been found that, while subjects perform simple tasks better in a group situation, this may not be so when a difficult task is involved.

To test this theory, modify the experiment by asking subjects to bounce a tennis ball, hitting it back down to the floor, at the top of each bounce, using the palm of the hand.

Count the number of bounces that they can make without any mistakes. You may give your subjects 1 minute or so to practise before commencing their trial. Follow through a similar analysis to that outlined in this section but modify H_1 to suit the theory which you are testing, i.e. with a difficult task subjects perform better on their own than in a group.

Section **10.10**   **Comparison of Paired Samples Using the Paired-sample t Test (Small Samples)**

Practical 10.15   **Variation in Weight**

Problem: **Does a person's weight differ between morning and evening?**

Method

Ask a sample of people, selected randomly, to weigh themselves last thing at night before they go to bed, and again first thing in the morning. On both occasions, they should wear the same clothing.

If members of your sample are unwilling to reveal their actual weights, that does not matter. It is sufficient for them to tell you whether they weighed more or less, and by how much.

The Null Hypothesis

We assume that the difference in weight (morning weight minus evening weight) follows a normal distribution $N(\mu, \sigma^2)$. Our null hypothesis is that there is no real difference in weights for evening and morning weighings so that the expected difference is 0:

$H_0: \mu = 0$

You could adopt the alternative hypothesis

$H_1: \mu < 0$

and use a one-tailed test, as it has been suggested that people generally do weigh less early in the morning than at any other time during the day.

The Test Statistic

We calculate $\bar{d}$, the mean of all the differences:

$$\bar{d} = \frac{\sum_i (x_{1i} - x_{2i})}{n}$$

Under H_0, $\bar{d}$ will have a $N(0, \sigma^2/n)$ distribution. Since we do not know σ^2, we estimate it using s^2, and $\bar{d}/(s/\sqrt{n})$ will follow a Student t distribution with $n-1$ degrees of freedom:

$$t_{n-1} = \frac{\bar{d}}{\sqrt{s^2/n}}$$

By working with individual differences, the effect caused by differences between subjects is reduced and the effect caused by the treatment is highlighted.

Our Results

Our results are given in Table 10.8.

Table 10.8   Masses in the morning and evening (to nearest 0.5 kg)

Subject	Mass (kg)		Difference (kg)	d^2
	Evening	Morning		
A	70.5	70.0	−0.5	0.25
B	61.0	60.5	−0.5	0.25
C	27.5	27.0	−0.5	0.25
D	55.0	55.5	+0.5	0.25
E	65.0	65.0	0	0
F	64.0	62.5	−1.5	2.25
G	50.0	49.0	−1.0	1.00
H	31.0	30.5	−0.5	0.25
I	45.5	44.5	0	0
J	41.0	40.0	−1.0	1.00

$$\bar{d} = \frac{-5.10}{10} = -0.5$$

$$s^2 = \frac{n\sum d^2 - \left(\sum d\right)^2}{n(n-1)}$$

$$= \frac{(10 \times 5.5) - (5)^2}{10 \times 9}$$

$$= \frac{55 - 25}{90}$$

$$= \frac{30}{90}$$

$$\approx 0.3333$$

The Significance Test

$H_0: \bar{d} = 0$

$H_1: \bar{d} < 0$

Working with a one-tailed t test, at the 5% level of significance, the critical value for t_9 is $t < -1.833$,

$$t_9 = \frac{-0.5}{\sqrt{0.3333/10}}$$

$$= \frac{-0.5}{\sqrt{0.03333}}$$

$$\approx \frac{-0.5}{0.1826}$$

$$\approx -2.738$$

The results obtained from our small sample of 10 subjects were significant, i.e. on the basis of our study we would have to reject H_0 and conclude that, at the 5% level, there was a difference between the weights of our subjects in the evening and the following morning.

If you look at our results in Table 10.8, you will notice that there is a great deal of variability between the subjects (some are adults and some are children). If we were to test between their overall mean weight in the morning and their overall mean weight in the evening (as in Section 10.9), the difference in means would not be significant owing to the high values of the sample variances. By using a paired sample test, we eliminate any effect of variability between subjects and concentrate on the effect of the experiment on each individual.

Section 10.11   The F Ratio Test

Practical 10.16   Variances

Problem: Do two samples have equal variances?

The F test is designed to test whether the variances calculated from two samples may be taken as coming from normal distributions with the same variance. We assume that the samples are independent (e.g. not paired).

Methods

Many of the studies outlined in this chapter yield data which can be usefully analysed using the F test. In particular, if two sample variances are to be combined to give a pooled two-sample estimate of the population variance, their compatibility can be checked by use of the F test.

The following sections include experiments which result in measurements being taken for two independent samples:

Sections 10.5 and 10.6   Comparison of male and female estimates of 1 min.
Section 10.7   Comparison of reaction times for two groups.
Section 10.9   Comparison of task performance by students working in groups or alone.

The Null Hypothesis

$H_0: \sigma_1^2 = \sigma_2^2 = \sigma^2$ (population variance)

$H_1: \sigma_1^2 \neq \sigma_2^2$

This is usually a two-tailed test.

The Test Statistic

$$F_{v_1 v_2} = \frac{\text{larger variance}}{\text{smaller variance}}$$

F is calculated with the larger variance divided by the smaller, so that the value of the ratio is always greater than 1. For both samples the variance should be calculated as

$$s^2 = \frac{\sum(x - \bar{x})^2}{n - 1}$$

or an equivalent formula such as

$$\frac{n\sum x^2 - \left(\sum x\right)^2}{n(n - 1)}$$

To look up the critical value of F in tables, values for the degrees of freedom are needed:

$v_1 = n_1 - 1$ from the first sample

$v_2 = n_2 - 1$ from the second sample

Tables are constructed for one-tailed tests; so, to find the critical value of F at the 5% level, look up the value at $2\frac{1}{2}$% if you require a two-tailed test.

Our Results

Our results (from Section 10.9) are given in Table 10.9.

Table 10.9 Comparison of task performance by students working in groups and alone

	s^2	n	v
Alone	2045.1	8	7
Group	7930.3	8	7

The Significance Test

$$F_{7,7} = \frac{7930.3}{2045.1} \approx 3.88$$

As we have two very small samples, we do require a very high value of F in order to achieve a significant result. The critical value of F for a two-tailed test at 5% level is

$$F_{7,7} = 4.99$$

Our result of 3.88 is not significant at the 5% level. However, it is a high value for F and would be significant at the 10% level on a two-tailed test. If we were to combine the results from the two samples to give a pooled estimate of the population variance, we would treat any further decision with some care.

Section 10.12 Non-parametric Tests for Small Samples: Introduction

In order to use a Student t test, for paired or independent samples, we must have measurements taken from a population which can be modelled by a normal distribution.

In many practicals, we cannot assume that this condition has been fulfilled.

If the population being sampled is not normal but a *t* test is used nonetheless, the probability of committing a type 1 error may be quite different from the nominal level of significance.

The discrepancy may be quite small for a distribution which is just slightly off normal but, for a uniform or U-shaped distribution, the problem will increase according to the degree of departure from normality.

In order to test hypotheses in these situations, non-parametric statistics have been developed. These tests make no assumptions at all about the distribution of the parent population.

If large samples are taken, then of course these problems do not arise, as $\bar{X} \sim N(\mu, \sigma^2/n)$ for large samples taken from any parent population with mean μ and variance σ^2.

Section 10.13   Wilcoxon Signed-rank Test Used to Test a Treatment on Paired Samples

Practical 10.17   Coaching in Table Tennis

Problem: **Does intensive coaching improve a player's skill at forehand drive in table tennis?**

This experiment can be adapted to test skill acquisition in other sports or in other areas outside sport as illustrated by the alternative practicals later in this section. The experiment has implications for sports coaching at all levels.

Equipment

You require the following.

A table tennis table.
A table tennis bat.
1 roll of masking tape.
A blackboard on a stand.
Trained 'feeders'.

Your subjects should be a group of children who have had no coaching in table tennis and who are not players.

Set up your equipment as in Figure 10.7.

Feeders

Feeders must be trained to bounce the ball into the target on their side of the table so that it bounces straight over the net into the quarter of the table on the player's right. (Note that we are coaching straightforward forehand drive and not return of serve.) The feeder must always stand in the same place and be trained to feed balls with consistency and accuracy.

Targets

The target for the feeder is a square with a side of 30 cm taped on the table as shown in the diagram. The target for the player is a large board placed to 'catch' any ball which is returned onto the correct section of the table (see diagram).

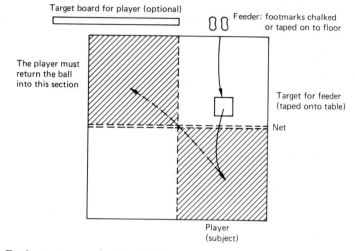

Target board for player (optional)

Feeder: footmarks chalked
or taped on to floor

The player must
return the ball
into this section

Target for feeder
(taped onto table)

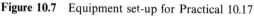

Net

Player
(subject)

Figure 10.7   Equipment set-up for Practical 10.17

Method

Pre-test

Your subjects should be pre-tested by feeding them 20 balls which they must try to return correctly. Count the number of successful returns.

Coaching

Give the children a period of practice, say 20 min on one day (or possibly 10 min on two or three separate days, provided that each child is given the same coaching pattern). You may like to extend this practical to see which pattern of coaching is most effective.

Alternatively the practical might be extended to include a *control* group who are given no specific training but allowed to play table tennis (or try to!) for an equivalent length of time.

Post-test

Retest the children with 20 balls after the coaching period. You may give them a rest first. Count the number of successful returns.

The Null Hypothesis

The null hypothesis is that the pupils' skill does not improve after a period of coaching, i.e. the improvements or positive differences in performance made by some children will be counterbalanced by the negative differences in scores for other children.

The Test Statistic

The Wilcoxon signed-rank test of differences is a non-parametric test. The changes in score (between a pre-test and post-test) are given ranks according to their absolute magnitude. Irrespective of whether it is positive or negative, the smallest difference is assigned the rank 1 and the next smallest the rank 2 and so on. *Zero differences are ignored.* The total of the positive ranks is then compared with the total of the negative ranks. The smaller of these two quantities is called T. In this example, where we expect a positive improvement under H_1, the total of the negative ranks should be the smaller quantity.

We can then calculate the probability that T is this value or less. If that probability is greater than 0.05, we can retain H_0. If it is less than 0.05, we can reject H_0 and can conclude that the children have made a significant improvement. We are conducting a one-tailed test in this case.

Our Results

10 children aged 11–12 years were used as subjects in our experiment. They were each given a pre-test of 20 forehand shots. They then had an intensive training period of 20 min working with one other child, a trained feeder and a coach. After a rest of 5 min, they were given a post-test of 20 shots. Their scores are given in Table 10.10.

Table 10.10 Pre-test and post-test scores

Child	Pre-test score	Post-test score
A	6	9
B	5	12
C	3	9
D	4	9
E	2	3
F	1	1
G	3	2
H	8	12
I	6	9
J	12	10

As child F has the same score before and after coaching, his results are excluded from the analysis.

The Significance Test

Firstly, calculate the differences in scores and rank them in terms of the absolute value (Table 10.11).

Table 10.11 Ranking of test scores

Child	Pre-test score x_1	Post-test score x_2	Difference $x_2 - x_1$	Rank of absolute difference
A	6	9	+3	4.5 (+)
B	5	12	+7	9 (+)
C	3	9	+6	8 (+)
D	4	9	+5	7 (+)
E	2	3	+1	1.5 (+)
F	1	1	0	—
G	3	2	−1	1.5 (−)
H	8	12	+4	6 (+)
I	6	9	+3	4.5 (+)
J	12	10	−2	3 (−)

E and G have differences of +1 and −1, but their absolute value is the same. In this case, we have tied ranks, and E and G must both be awarded the rank 1.5. The next largest difference is J with a difference of −2, and so J is awarded the rank 3, and so on.

The sum of the positive ranks is

$$4.5 + 9 + 8 + 7 + 1.5 + 6 + 4.5 = 40.5$$

The sum of the negative ranks is

$$1.5 + 3 = 4.5$$

As a check the total for all the ranks should be $n(n+1)/2$. Here $n = 9$. The total for all ranks is $(1/2) \times 9 \times 10 = 45$. The lower total is for the negative ranks; so $T = 4.5$.

Now we have to find the probability of finding a total for T of 4.5 or less. There are 9 ranks in all, each of which could be positive or negative; so we have $2^9 = 512$ different outcomes possible.

We can then compute the number of these outcomes which would give us a total for T of 4.5 or lower.

	1.5			4.5						
Rank	1	2	3	4	5	6	7	8	9	
	+	+	+	+	+	+	+	+	+	$T = 0$
	−	+	+	+	+	+	+	+	+	$T = 1.5$
	+	−	+	+	+	+	+	+	+	$T = 1.5$
	+	+	−	+	+	+	+	+	+	$T = 3$
	−	−	+	+	+	+	+	+	+	$T = 3$
	+	+	+	−	+	+	+	+	+	$T = 4.5$
	−	+	−	+	+	+	+	+	+	$T = 4.5$
	+	−	−	+	+	+	+	+	+	$T = 4.5$
	+	+	+	+	−	+	+	+	+	$T = 4.5$

Any other arrangement of positive and negative signs will give a value of T equal to 6 or greater.

So we have 9 possible outcomes in which T for the negative ranks has a total of 4.5 or less. Thus,

$$P(T \leqslant 4.5) = \frac{9}{512} \approx 0.018$$

Our results are significant at the 5% level, and we can conclude that this group of children did make significant improvements after their coaching.

Section 10.14   'Practice makes Perfect'

The practicals in this section should all show an improvement in task performance after a period of practice.

Practical 10.18   Reactions II

Test a group of subjects with the 'reaction timer', timing them before and after a period of practice (see Section 10.7, Practical 10.8).

Practical 10.19   Sorting

Mix the following objects on a tray: 4 1 p pieces, 4 5 p pieces, 4 20 p pieces, 4 buttons and 4 washers. Blindfold your subject and measure the time taken to sort the objects into five piles of like objects. Repeat the experiment again, and their second attempt should be faster. Use about 10 subjects in all, but do not let them watch each other (as they will learn while they are watching).

Practical 10.20   Tracing

Print a simple shape six times onto a sheet of paper. Ask your subjects to draw round the shape using their left (or non-dominant) hand. Time their first and last tracing. A page of shapes for use in this experiment is printed on a photocopiable page (page PC13). You should find their sixth try is faster than their first.

Section 10.15   Comparison of Samples

Practical 10.21   Flame-retardant Sprays

Problem: **Does a flame-retardant spray make fabrics burn more slowly?**

This practical should be conducted in a laboratory or in the open air with a fire extinguisher available for use if needed.

Materials

You require a flame-retardant spray. These are not widely available, but your local fire prevention officer at the Fire Brigade offices can advise you to where you can purchase one. (We were directed to a local theatrical supplier.) You will also need small pieces of material about 10 cm × 10 cm, *two* of each type to be tested. You may use all the furnishing fabrics of the types used for curtains and furniture coverings, or all clothing fabrics, or some of each. You should find out the fibre content of each fabric; wool, cotton, linen, rayon, terylene, nylon, etc.

Spray *one* sample of each type according to the instructions on the can. Allow the spray to dry thoroughly. Then test the two samples of each fabric by recording the time taken for each to burn.

The Null Hypothesis

Your null hypothesis is that the flame-retardant spray makes no difference to the time taken for material to burn.

Analysis of Results

Your analysis will follow the method set out in Section 10.13.

Variation of this Practical

Another variation of this practical would be to test the effect of fire-resistant paint on the burning times of different types of wood. Again, your local fire prevention officer will tell you where to buy fire-resistant paint and this may be more readily available than the spray.

Section 10.16   Comparison of Paired Samples Using the Sign Test

Practical 10.22   Alternative Analysis of Practicals 10.17–10.20

Practicals 10.17–10.20 yield data which can be analysed using the sign test.

The sign test is simpler than the Wilcoxon signed-rank test. It is a non-parametric test which may be used to test the differences between paired results. It makes no assumptions about the population distribution from which the sample readings were taken.

When the sign test is used, the difference between two sets of measures (pre-treatment and post-treatment) are analysed purely in terms of whether they are positive or negative. Differences are not ranked according to absolute magnitude.

Problem

We shall use the sign test here to test the hypothesis from the experiment outlined in Section 10.13. Pupils were tested before and after a coaching session aimed at improving their performance at forehand drive in table tennis. On both occasions the children were given 20 shots at the ball.

The Null Hypothesis

The null hypothesis is that there is no improvement in the children's scores. As the scores may not come from a normal distribution, we cannot make any assumptions about the mean of the differences, but we can say that the median of the differences should equal zero. Also under this null hypothesis, a positive improvement or a negative deterioration are both equally likely, i.e.

$$P(+ \text{ difference}) = 1/2$$

$$P(- \text{ difference}) = 1/2$$

We rule out any cases where there is zero difference. Our alternative hypothesis is that pupils do improve with practice, i.e. $P(+ \text{ difference}) > 1/2$. If there are n pupils, then the number of positive differences can be modelled by the binomial distribution $B(n, 1/2)$.

Our Results

Our results are given in Table 10.12.

Table 10.12 Signs of difference

Pupil	Pre-test score	Post-test score	Sign of difference
A	6	9	+
B	5	12	+
C	3	9	+
D	4	9	+
E	2	3	+
F	1	1	Zero
G	3	2	−
H	8	12	+
I	6	9	+
J	12	10	−

In our experiment, nine non-zero differences were recorded. Thus, under H_0, the distribution of positive (or negative) differences should follow a binomial distribution $B(9, 1/2)$.

The Test Statistic

The test statistic is whichever is the smaller of the number of positive differences or the number of negative differences. Here we have two negative differences and we wish to calculate the probability of finding two or less.

The Significance Test

We can calculate the probability of each possible number of negative differences using the binomial distribution $B(9, 1/2)$ which is our mathematical model for the situation under the null hypothesis (Table 10.13).

Table 10.13 Probabilities

Number x of negative differences	$P(x)$	
0	$\left(\dfrac{1}{2}\right)^9$	≈ 0.00195
1	$9 \times \left(\dfrac{1}{2}\right)^9$	≈ 0.01758
2	$\dfrac{9 \times 8}{2 \times 1} \times \left(\dfrac{1}{2}\right)^9$	≈ 0.0703
3	$\dfrac{9 \times 8 \times 7}{3 \times 2 \times 1} \times \left(\dfrac{1}{2}\right)^9$	≈ 0.1641
4	$\dfrac{9 \times 8 \times 7 \times 6}{4 \times 3 \times 2 \times 1} \times \left(\dfrac{1}{2}\right)^9$	≈ 0.2461
5	$\dfrac{9 \times 8 \times 7 \times 6}{4 \times 3 \times 2 \times 1} \times \left(\dfrac{1}{2}\right)^9$	≈ 0.2461
6	$\dfrac{9 \times 8 \times 7}{3 \times 2 \times 1} \times \left(\dfrac{1}{2}\right)^9$	≈ 0.1641
7	$\dfrac{9 \times 8}{2 \times 1} \times \left(\dfrac{1}{2}\right)^9$	≈ 0.0703
8	$9 \times \left(\dfrac{1}{2}\right)^9$	≈ 0.01758
9	$\left(\dfrac{1}{2}\right)^9$	≈ 0.00195

We are conducting a one-tailed test in which we shall want to reject H_0 if we have a very low number of negative differences. The critical region should include a total probability of 0.05 or less.

$P(x \leqslant 1) = 0.0195$ just below 2%

$P(x \leqslant 2) = 0.0898$ just below 9%

Alternatively a critical region of $x \leqslant 2$ will give us a test of size 9% (Figure 10.8).

At the 5% (and also 2%) level, we reject H_0 if we have no negative differences or one negative difference; otherwise, we retain H_0.

We have two negative differences; so we retain H_0 and conclude that at the 5% level the coaching has not brought about any more improvements than we could expect by chance.

This conclusion is different to the one we arrived at using the Wilcoxon signed-rank test (used in Section 10.11) which takes the magnitude of the differences into account.

We could also, of course, conclude that at the 10% level, coaching has improved performance which agrees with our previous conclusion but at a different significance level.

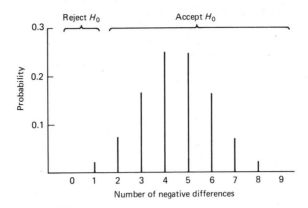

Figure 10.8 Binomial probabilities for the sign test

Section 10.17 Comparison of Two Distributions Using the Mann–Witney *U* Test

Practical **10.23 Testing of a New Teaching Method**

Problem: **Are two groups of pupils, to be used to compare two teaching methods, comparable at the beginning of the experiment?**

A local school wishing to test a new method of French teaching tested two classes at the beginning of their experiment. One class was to be taught a topic by the new method, and the other by a traditional method. At the end of the topic, both classes were tested again. The problem here is to determine whether the children in those classes were reasonably comparable at the beginning of their courses. (The full results are included in Table 10.16 so that you can conduct further tests yourself to determine which method of teaching was more effective.)

The Mann–Witney *U* test will detect differences in the overall distributions rather than differences in the distribution means.

The Null Hypothesis

The null hypothesis is that the marks of the two groups come from the same distribution.

Our Results

Table 10.14 shows our results.

The Test Procedure

1. Rank all the scores as though they are in one group, giving rank 1 to the lowest score.
2. In the case of tied ranks, award the average of those ranks.
3. Sum the ranks for the groups separately, to obtain R_1 and R_2.
4. Define

$$U_1 = n_1 n_2 + \frac{n_1(n_1 + 1)}{2} - R_1$$

$$U_2 = n_1 n_2 + \frac{n_2(n_2 + 1)}{2} - R_2$$

where n_1 and n_2 are the sizes of samples 1 and 2, respectively.
5. Calculate U_1 and U_2.

Table 10.14

Pre-test scores		Pre-test scores *(continued)*	
Group 1	Group 2	Group 1	Group 2
11	13	12	10
13	19	12	17
11	9	6	13
9	10	11	20
10	18	11	19
25	19	24	13
18	10	13	29
14	19	19	15
27	5	9	17
23	29	10	11
19	19		10
10	16		8
9	8		
8	30	$n = 24$	$n = 26$

As a check, the sum of all the ranks can be calculated: $R_1 + R_2 = N(N+1)/2$, where N is the total number of observations of $n_1 + n_2$.

Also $U_1 + U_2 = n_1 n_2$.

Note that high values of R_1 (and hence low values of U_1) would indicate high performance by group 1. For small samples, U_1 or U_2 can be calculated directly. U_1 is obtained by arranging the $n_1 + n_2$ observations in order of size and counting the number of times that an item from the first sample precedes one from the second sample, e.g. (denoting by A an item from the first sample and by B an item from the second sample and putting in order of size) ABBBAB gives $U_1 = 4 + 1 = 5$ ($U_2 = 1 + 1 + 1 = 3$). If both samples are less than 20, consult U tables. If at least one sample contains more than 20 values, Z may be computed from the following formula and normal tables used:

$$z = \frac{U - n_1 n_2/2}{\sqrt{[n_1 n_2 (n_1 + n_2) + 1]/12}}$$

U_1 or U_2 may be used—the answers will give a total probability of 1.

The Significance Test

The significance test results are given in Table 10.15.

Check for sum of ranks

$$R_1 + R_2 = 564 + 711 = 1275$$

$$\frac{N(N+1)}{2} = \frac{50 \times 51}{2} = 1275$$

$$U_1 = n_1 n_2 + \frac{n_1(n_1 + 1)}{2} - R_1$$

$$= 24 \times 26 + \frac{24 \times 25}{2} - 564$$

$$= 624 + 300 - 564$$

$$= 360$$

Table 10.15   Results

Pre-test score, group 1	Rank	Pre-test score, group 2	Rank
11	19	13	26
13	26	19	39
11	19	9	7.5
9	7.5	10	13
10	13	18	34.5
25	46	19	39
18	34.5	10	13
14	29	19	39
27	47	5	1
23	44	29	48.5
19	39	19	39
10	13	16	31
9	7.5	8	4
8	4	30	50
12	22.5	10	13
12	22.5	17	32.5
6	2	13	26
11	19	20	43
11	19	19	39
24	45	13	26
13	26	29	48.5
19	39	15	30
9	7.5	17	32.5
10	13	11	19
		10	13
		8	4
$n_1 = 24$	$R_1 = 564$	$n_2 = 26$	$R_2 = 711$

$$U_2 = n_1 n_2 + \frac{n_2(n_2 + 1)}{2} - R_2$$

$$= 24 \times 26 + \frac{26 \times 27}{2} - 711$$

$$= 624 + 351 - 711$$

$$= 264$$

Check for U_1 and U_2

$$U_1 + U_2 = 360 + 264 = 624 (= 24 \times 26 = n_1 n_2)$$

The Significance Test

H_0: the two samples come from the same distribution

H_1: the two samples come from different distributions

This is a two-tailed test.

The critical value of Z is $|Z| > 1.96$.

$$z = \frac{U_1 - n_1 n_2 / 2}{\sqrt{n_1 n_2 (n_1 + n_2 + 1)/12}}$$

$$= \frac{360 - 624/2}{\sqrt{624 \times 51/12}}$$

$$= \frac{360 - 312}{\sqrt{2652}}$$

$$\approx \frac{48}{51.50}$$

$$\approx 0.932$$

This value of z is not significant at the 5% level and we conclude that, at this level, there was no difference between the two groups of children at the beginning of the experiment (Table 10.16).

Table 10.16 Results for Practical 10.23 using different teaching methods (children in group 1 were taught a topic in French by the traditional method, while children in group 2 were taught that topic by a new method; their test scores are given before and after that part of their course)

Group 1 scores		Group 2 scores	
Pre-test	Post-test	Pre-test	Post-test
11	22	13	17
13	28	19	20
11	27	18	13
9	37	19	19
10	35	10	12
25	42	19	18
18	34	5	10
14	27	29	23
27	45	19	15
23	35	16	18
19	48	8	12
10	31	30	24
9	42	10	12
8	48	17	22
12	44	13	16
12	48	20	20
6	24	19	16
11	21	13	19
11	43	29	31
24	44	15	18
13	37	17	17
19	43	11	20
9	34	10	9
10	37	16	11
		10	9
		8	15

Acknowledgements

John Alderson and Celia Brackenridge, both of Sheffield City Polytechnic, provided ideas and help with the physical education and sports experiments in this and Chapter 11. Some of the practicals entailing psychology were suggested by John Elliott of the University of Sheffield.

Chapter 11 Correlation

Introduction

Do tall parents produce tall children?

Is a person's weight related to his or her height?

Is there a relationship between rainfall and hours of sunshine?

Is the amount of money spent by pupils at the school tuck-shop related to their age?

In this chapter, we look for a way to examine the possible relationship between two measurements so that we can answer questions such as these. If we can develop a measure of the strength of a relationship, we can solve such problems as which is the better predictor of the amount of money spent by pupils at the tuck-shop, their age or the amount of pocket money they receive.

Section 11.1 Practicals

Choose from the following list of practicals, but try to ensure that—among the class as a whole—at least one practical from each group has been attempted.

For any of the following suggestions, you are required to record two measurements (or results) for each member of your sample.

Group A

Practical 11.1 Lengths and Widths of Leaves

For a random sample of 15 leaves (of the same species) measure the length and width of each leaf. You may have these data already available from your practical work for Chapter 5.

Practical 11.2 Heights and Weights of Students

Measure and record *either* of the following.

1. The heights and weights of 15 students of the same sex.
2. The heights and ages for a sample of younger children of various ages.

Practical 11.3 Heights of Students and their Parents

Measure the heights of 15 students of the same sex and find also *either* of the following.

1. The heights of their fathers.
2. The heights of their mothers.

Practical 11.4 School Tuck-shop

Conduct a survey at your school tuck-shop. For each customer, note down the following.

1. The amount of money that they spend.
2. Their age (or year group in the school).
3. Ask them the amount of pocket money that they receive each week.

If you collect up a large number of results you may need to take a random sample of about 20 of all your replies. As you work through the chapter, you will be able to investigate possible relationships between 1 and 2, between 1 and 3, or between 2 and 3.

Group **B**

Practical **11.5   Weights of Coins**

Weigh 15 coins of the same denomination (2 p, 10 p or 50 p are suggested) and record their weights and ages. An electronic balance accurate to within 0.01 g is needed for this practical.

Practical **11.6   Daily Rainfall and Sunshine Data**

Obtain data for daily rainfall and hours of sunshine from your local weather station for a period of at least 2 weeks.

Group **C**

Practical **11.7   First Names and Surnames**

Do parents with long surnames give their children short first names? For a group of at least 15 people, count the number of letters in their first names and their surnames (use the maiden name for married women).

Practical **11.8   Daily Maximum and Minimum Temperatures**

Obtain figures for daily maximum and minimum temperatures from your local weather station (or your school's geography department). You require figures for at least 2 weeks.

Practical **11.9   Lung Capacities**

In order to join the Fire Brigade, applicants must be able to breathe in and to expand their chest to a measurement of at least 40 inches. The theory behind this is that fire fighters need large lung capacities. Ask 15 people to breathe in and record their expanded chest measurement. Also record the length of time in seconds for which they are able to hold their breath.

Section **11.2   Our Results**

Practical 11.1

Table 11.1 shows the measurements for ten laurel leaves.

Table 11.1   Lengths and widths of laurel leaves

Length (mm)	103	119	146	148	87	134	151	147	122	129
Width (mm)	35	37	41	53	35	39	50	42	33	43

At a first inspection, these data seem to suggest that the longer leaves also tend to be wider.
This can be seen more clearly if we illustrate the data on a scatter diagram (Figure 11.1). Length in millimetres is shown on the *x* axis and width in millimetres on the *y* axis so that each leaf is represented on the diagram by a single point.

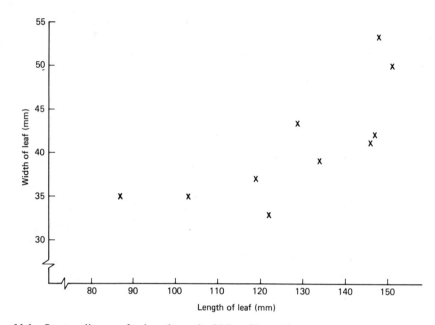

Figure 11.1 Scatter diagram for lengths and widths of laurel leaves

The points do not form a clearly defined line but, although there is considerable variation between the points, they do not appear to be entirely randomly placed either.

We can investigate whether leaves which are above average in length are also above average in width. The mean length is 128.6 mm and the mean width is 40.8 mm. When lines are drawn on the scatter diagram (Figure 11.2) to show these mean measurements, it is clear that above-average lengths and widths do go together as do below-average lengths and widths. All the points except one fall into the top right or bottom left quadrants on

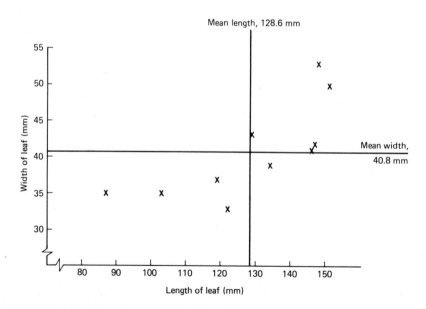

Figure 11.2 Mean length and mean width of laurel leaves

the diagram. Length and width appear to vary together, i.e. there appears to be a positive relationship between them.

Practical 11.5

Table 11.2 shows the ages of ten 10 p coins together with their masses measured in grams.

Table 11.2   Masses and ages of 10 p coins

Age (years)	17	18	7	13	12	10	20	18	25	12
Mass (g)	11.20	11.20	11.19	11.22	11.36	11.31	11.00	11.09	10.93	11.25

Any possible relationship between age and mass can be seen more clearly from the scatter diagram (Figure 11.3) drawn by plotting age as the x coordinate and mass as the y coordinate for each coin.

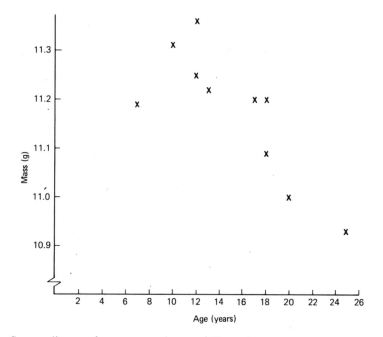

Figure 11.3   Scatter diagram for masses and ages of 10 p coins

This diagram seems to suggest that older coins generally weigh less than newer ones. If the mean age and mean mass are calculated (15.2 years and 11.175 g, respectively) and lines drawn on the scatter diagram (Figure 11.4), this relationship becomes more apparent. Coins with below-average ages do have above-average masses, and vice versa. Only two coins lie outside the top left and bottom right quadrants.

This example appears to show a different kind of relationship between x and y than that shown in the first example. As age increases, mass decreases. This type of relationship is called a negative relationship.

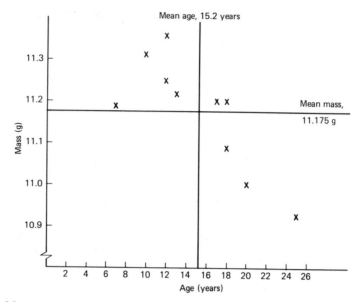

Figure 11.4   Mean mass and mean age of 10 p coins

Practical Follow-up **11.1**

For each practical that you have completed, draw a scatter diagram for each set of measurements and draw lines onto the diagram as shown in Section 11.2. Do your measurements show a positive relationship or a negative relationship are the points on the diagram essentially random, suggesting that no general relationship exists between the two variables?

Section **11.3**   **A Measure of the Strength of a Relationship**

Looking at our Practical 11.5 results again, we shall try to derive a statistical measure for the strength of the relationship between a coin's age and its mass.

For each coin, we can calculate the difference from the mean of both its age and its mass (Table 11.3).

Table 11.3   Differences from the mean age and from the mean mass

Age x of coin (years)	Mass y of coin (g)	Difference $x - \bar{x}$ from mean age (years)	Difference $y - \bar{y}$ from mean mass (g)
17	11.20	1.8	0.025
18	11.20	2.8	0.025
7	11.19	−8.2	0.015
13	11.22	−2.2	0.045
12	11.36	−3.2	0.185
10	11.31	−5.2	0.135
20	11.00	4.8	−0.175
18	11.09	2.8	−0.085
25	10.93	9.8	−0.245
12	11.25	−3.2	0.075

Note that a positive difference $x - \bar{x}$ is generally associated with a negative difference $y - \bar{y}$ and vice versa (only the first two coins are exceptions).

If we then multiply $x - \bar{x}$ by $y - \bar{y}$, most points will yield a negative product, so that for a negative relationship $\sum[(x - \bar{x})(y - \bar{y})]$ will be negative (Table 11.4).

Table 11.4 The products $(x - \bar{x})(y - \bar{y})$

Age x (years)	Mass y (kg)	Deviation $x - \bar{x}$ of age from mean (years)	Deviation $y - \bar{y}$ of mass from mean (kg)	$(x - \bar{x})(y - \bar{y})$
17	11.20	1.8	0.025	0.045
18	11.20	2.8	0.025	0.07
7	11.19	−8.2	0.025	−0.123
13	11.22	−2.2	0.045	−0.088
12	11.36	−3.2	0.185	−0.592
10	11.31	−5.2	0.135	−0.702
20	11.00	4.8	−0.175	−0.84
18	11.09	2.8	−0.085	−0.238
25	10.93	9.8	−0.245	−2.401
12	11.25	−3.2	0.075	−0.24

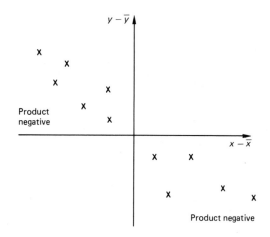

Figure 11.5 Negative relationship, i.e. negative product

For a negative relationship, $\sum[(x - \bar{x})(y - \bar{y})]$ is negative (Figure 11.5).

For a positive relationship, however $\sum[(x - \bar{x})(y - \bar{y})]$ will be positive (Figure 11.6).

Practical Follow-up 11.2

Calculate $\sum[(x - \bar{x})(y - \bar{y})]$ for each pair of variables for the practicals that you have completed.

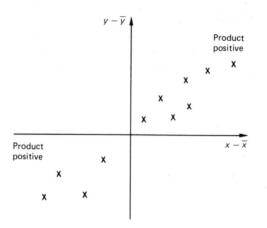

Figure 11.6 Positive relationship, i.e. positive product

Section **11.4** **The Correlation Coefficient**

Although the sign of $\sum[(x-\bar{x})(y-\bar{y})]$ gives us an indication of the type of relationship (positive or negative) between x and y, the absolute magnitude of this total depends both on the units of measurement used and on the size of the sample.

A more useful measure would be one which was independent of the sample size and the scales of the measurement used. We can allow for the sample size by dividing by n to find the mean value of $(x_i-\bar{x})(y_i-\bar{y})$. The measure

$$\frac{1}{n}\sum_i[(x_i-\bar{x})(y_i-\bar{y})]$$

is called the sample covariance.

Also, we can allow for the different scales of measurement by measuring the deviations from the means of x and y in terms of the sample standard deviation (SD) for each variable. We have

$$SD(x)=\sqrt{\frac{\sum_i(x_i-\bar{x})^2}{n}}$$

and

$$SD(y)=\sqrt{\frac{\sum_i(y_i-\bar{y})^2}{n}}$$

Our standardised measure becomes

$$\frac{1}{n}\sum_i\frac{x_i-\bar{x}}{SD(x)}\frac{y_i-\bar{y}}{SD(y)}$$

and, as the n values cancel out, this simplifies to

$$\frac{\sum_{i=1}^{n}(x_i-\bar{x})(y_i-\bar{y})}{\sqrt{\sum_{i=1}^{n}(x_i-\bar{x})^2\sum_{i=1}^{n}(y_i-\bar{y})^2}}$$

This is known as the product moment correlation coefficient.

This formula is generally given in the following form which is easier to use for calculation purposes:

product moment correlation coefficient $r_{xy} =$

$$\frac{n \sum_{i=1}^{n} x_i y_i - \sum_{i=1}^{n} x_i \sum_{i=1}^{n} y_i}{\sqrt{\left[n \sum_{i=1}^{n} x_i^2 - \left(\sum_{i=1}^{n} x_i \right)^2 \right]\left[n \sum_{i=1}^{n} y_i^2 - \left(\sum_{i=1}^{n} y_i \right)^2 \right]}}$$

The proof of the equivalence of this formula is given at the end of this chapter.

A Worked Example

Here is the calculation of the product moment correlation coefficient for the lengths and widths of leaves (Practical 11.1). The necessary calculations are shown in full (Table 11.5 and below). You may have a calculator which will calculate the correlation coefficient directly from the values of (x_i, y_i). If not, you will have to follow our method of working, *being careful* not to round off any figures during the calculation (generally in the second line of working marked *).

Table 11.5 Calculation of product moment correlation coefficient

Length x (mm)	Width y (mm)	x^2	y^2	xy
103	35	10 609	1 225	3 605
119	37	14 161	1 369	4 403
146	41	21 316	1 681	5 986
148	53	21 904	2 809	7 844
87	35	7 569	1 225	3 045
134	39	17 956	1 521	5 226
151	50	22 801	2 500	7 550
147	42	21 609	1 764	6 174
122	33	14 884	1 089	4 026
129	43	16 641	1 849	5 547
1 286	408	169 450	17 032	53 406

$$r_{xy} = \frac{n \sum xy - \sum x \sum y}{\sqrt{\left[n \sum x^2 - \left(\sum x \right)^2 \right]\left[n \sum y^2 - \left(\sum y \right)^2 \right]}}$$

$$= \frac{10 \times 53\,406 - 1286 \times 408}{\sqrt{(10 \times 169\,450 - 1286^2)(10 \times 17\,032 - 408^2)}}$$

$$= \frac{534\,060 - 524\,688}{\sqrt{(1\,694\,500 - 1\,653\,796)(170\,320 - 166\,464)}} *$$

$$= \frac{9372}{\sqrt{40\,704 \times 3856}}$$

$$\approx \frac{9372}{12\,528.2}$$

$$\approx 0.748$$

We shall interpret this value in later sections.

Change in Origin and Units

The correlation coefficient is not affected by a change in origin or units during its calculation. For a second example, suppose that we used an origin of 100 mm for x (leaf lengths) and 40 mm for y (leaf widths), so that

$$u = x - 100$$

$$v = y - 40$$

Note that a change in origin gives us smaller numbers to work with and eliminates any risk of 'losing' significant figures in the early stages of the calculation (Table 11.6).

Table 11.6 Calculation of product moment correlation coefficient

Length (mm)	Width (mm)	$u = x - 100$	$v = y - 40$	u^2	v^2	uv
103	35	3	−5	9	25	−15
119	37	19	−3	361	9	−57
146	41	46	1	2 116	1	46
148	53	48	13	2 304	169	624
87	35	−13	−5	169	25	65
134	39	34	−1	1 156	1	−34
151	50	51	10	2 601	100	510
147	42	47	2	2 209	4	94
122	33	22	−7	484	49	−154
129	43	29	3	841	9	87
		286	8	12 250	392	1 166

$$r = \frac{n\sum uv - \sum u \sum v}{\sqrt{\left[n\sum u^2 - \left(\sum u\right)^2\right]\left[n\sum v^2 - \left(\sum v\right)^2\right]}}$$

$$= \frac{10 \times 1166 - 286 \times 8}{\sqrt{(10 \times 12\,250 - 286^2)(10 \times 392 - 8^2)}}$$

$$= \frac{116\,600 - 2288}{\sqrt{(122\,500 - 81\,796)(3920 - 64)}}$$

$$= \frac{9372}{\sqrt{40\,704 \times 3856}}$$

$$\approx \frac{9372}{12\,528}$$

$$\approx 0.748$$

At the end of this chapter, we give a proof of the fact that a change in origin and/or units does not affect the correlation coefficient.

Practical Follow-up 11.3

Calculate the product moment correlation coefficient for each set of data you have from your practicals.

Section **11.5   Interpretation of the Product Moment Correlation Coefficient**

The product moment correlation coefficient may take any value between -1 and $+1$ (the proof is given at the end of this chapter).

If $r_{xy} = +1$, there is a perfect positive linear relationship between x and y and all the points (x_i, y_i) lie exactly on a straight line with a positive gradient. However, if $r_{xy} = -1$, there is again a perfect linear relationship between x and y but on this occasion the straight line joining all the points (x_i, y_i) has a negative gradient.

A zero (or close to zero) correlation implies a random scattering of points with no linear relationship at all. However, this does not necessarily imply a total lack of any relationship between x and y, as points on a quadratic curve could produce a value of r_{xy} close to zero. Yet in this case there is a predictive relationship between x and y.

Figures 11.7, 11.8 and 11.9 show the sorts of scatter diagram which could give rise to various values of r_{xy}. Note that r_{xy} is not a measure of the slope of the line but rather the degree of linearity between points.

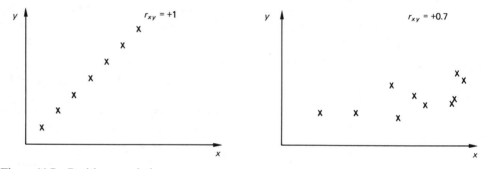

Figure 11.7   Positive correlation

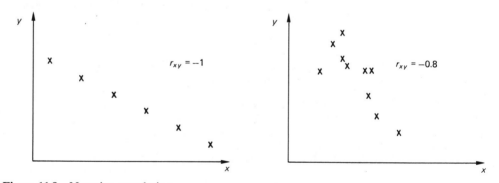

Figure 11.8   Negative correlation

Even a high correlation, either positive or negative, which shows a relationship which is approximately linear, does not imply cause and effect. For instance, there has over recent years been a steady rise in the number of burglaries, together with rising sales of video recorders. Which, if any, of the following explanations seem reasonable?

1. The increasing number of video recorders in affluent homes is a temptation to thieves.
2. The increase in sales of videos arises because people have to buy replacements for those which are stolen.
3. The two variables are unrelated.

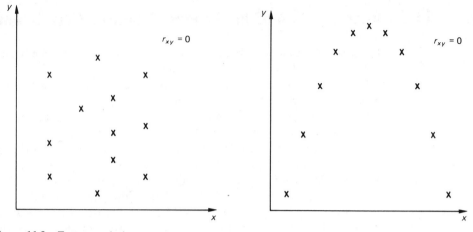

Figure 11.9 Zero correlation

4. The increase in both variables depends to some extent on a third variable which could be a general increase in population or people's rising expectations of their standard of living.

Certainly, any correlation coefficient should be interpreted with care. Even if a causal relationship does seem likely, the correlation coefficient will not, in itself, indicate which variable depends on which.

Such non-causal correlations are sometimes called *spurious correlations*. Other examples will arise where one of the two variates used is part of the other. For instance there will be a high correlation between male unemployment figures and total unemployment figures (for male and female together). In this case, we should be looking for any possible relationship between male unemployment and female unemployment. (Do both rise together? Or does female unemployment fall as male unemployment rises?)

Practical Follow-up 11.4

Compare the correlation coefficients obtained for all your practicals. Which pairs of variables appear to show the following.

1. A strong positive relationship.
2. A strong negative relationship.

Section 11.6 The Sampling Distribution of r_{xy}

So far, we have considered the calculation of a sample correlation coefficient from sample values of x and y. Each sample of a given size taken from a population will yield a value of r_{xy}. In the same way that sample means $\bar{x}$ calculated for all possible samples of size n form a sampling distribution with a mean μ, so all values of r_{xy} calculated for all possible samples will form a sampling distribution.

Each sample correlation coefficient can be taken as an estimate of the population (or distribution) correlation coefficient ρ (pronounced 'rho').

$$\rho = \frac{E\{[X - E(X)][Y - E(Y)]\}}{\sqrt{E[X - E(X)]^2 E[Y - E(Y)]^2}}$$

or

$$\rho = \frac{\text{Covar}(X, Y)}{\sqrt{\text{Var}(X)\,\text{Var}(Y)}}$$

where Covar is the covariance. If the sample is drawn at random from a distribution in which x and y have a joint normal distribution, then the sample correlation coefficient r_{xy} is an unbiased estimate of ρ.

The sampling distribution of r depends on the actual value of ρ. When $\rho = 0$ and x and y are normally distributed, the sampling distribution takes a simpler form which is symmetrical about 0. To test whether a sample correlation coefficient is significantly different from zero, tables may be used. In order to use these tables, the number of degrees of freedom must be calculated as $n - 2$ and significant values of r are shown in the table (for a two-tailed test). We 'lose' two degrees of freedom as $\bar{x}$ and $\bar{y}$ are estimated from the sample.

For example, our sample of 10 coins yielded a correlation coefficient of -0.7988 between the age and weight of the coins. Is there a significant relationship between age and weight, i.e. is r_{xy} significantly different from zero? For a sample of size $n = 10$ there are 8 degrees of freedom. By consulting tables, we find that r_{xy} must lie outside the range ± 0.632 to be significant at the 5% level (and outside the range ± 0.765 for 1% significance). Only 1% of all sample correlation coefficients calculated for samples of size 10 taken from a population in which x and y are totally uncorrelated will lie outside the range ± 0.765.

Our sample value of -0.7988 does lie outside this range; so we do have a correlation coefficient which is significant at the 1% level. There does appear to be a real relationship between a coin's age and its weight, with older coins weighing less.

A Confidence Interval for ρ

In order to construct confidence intervals or to carry out significance tests for values of $\rho \neq 0$,

$$Z' = \frac{1}{2}\ln\left(\frac{1+r}{1-r}\right)$$

is approximately normally distributed with mean

$$\frac{1}{2}\ln\left(\frac{1+\rho}{1-\rho}\right)$$

and variance

$$\frac{1}{n-3}$$

Tables are available to transform r to Z'.

Example

Our sample of 10 coins yielded a correlation coefficient $r_{xy} = -0.7988$. From this, we can calculate a 95% confidence interval for ρ. The transformed value of $r = -0.7988$ is $Z' = -1.099$ approximately. The variance of the sampling distribution is $1/(n-3) = 1/7$, giving a standard deviation of $1/\sqrt{7} = 0.378$. The confidence interval for Z' is $-1.099 \pm 1.96 \times 0.378 = -1.099 \pm 0.74088$, i.e. $-1.8399 < Z' < -0.35812$. By referring to tables, we can transform Z' back to r to give our confidence interval for ρ as $-0.9508 < \rho < -0.3434$.

Section 11.7 Bivariate Frequency Tables

Our Results for Practical 11.7

Suspecting that parents with long surnames might give their children short first names, we counted the number of letters in each for a group of 29 students (Table 11.7).

Table 11.7 Numbers of letters in surnames and first names

Number x of letters in surname	Number y of letters in first name	Number x of letters in surname (*continued*)	Number y of letters in first name (*continued*)
6	7	7	6
5	5	6	6
4	5	6	6
7	7	4	4
8	6	6	5
6	3	6	6
6	7	7	6
7	6	5	4
5	7	6	7
6	4	7	7
6	5	6	7
5	5	6	5
6	6	6	4
5	7	5	7
8	5		

These results can be summarised in a bivariate frequency table (Table 11.8) where each cell (or square) in the table contains the frequency for that pair of x, y values.

Table 11.8 Bivariate frequency table

y \ x	Number of letters in surname				
	4	5	6	7	8
3			1		
4	1	1	2		
5	1	2	3		1
6		0	4	2	1
7		3	4	3	

(Number of letters in first name)

In order to calculate the product moment correlation coefficient, the formula for r_{xy} is modified to include frequencies:

$$r_{xy} = \frac{\sum f \sum fxy - \sum fx \sum fy}{\sqrt{\left[\sum f \sum fx^2 - \left(\sum fx\right)^2\right]\left[\sum f \sum fy^2 - \left(\sum fy\right)^2\right]}}$$

It is also useful to set out the table with large squares so that each one may be used to calculate $\sum fxy$ by setting out the calculation thus:

Table 11.9 shows the calculation shown in full, although most calculators will do this for you.

Table 11.9 Calculation of r_{xy}

y \ x	3	4	5	6	7	8	f	fy	fy²	fxy
3				¹⁸1₁₈			1	3	9	18
4		¹⁶1₁₆	²⁰1₂₀	²⁴2₄₈	·		4	16	64	84
5		²⁰1₂₀	²⁵2₃₀	³⁰3₉₀		⁴⁰1₄₀	7	35	175	200
6				³⁶4₁₄₄	⁴²2₈₄	⁴⁸1₄₈	7	42	252	276
7			³⁵3₁₀₅	⁴²4₁₆₈	⁴⁹3₁₄₇		10	70	490	420
f		2	6	14	5	2	29	166	990	998
fx		8	30	84	35	16	173			
fx²		32	150	504	245	128	1059			
fxy		36	175	468	231	88	998			

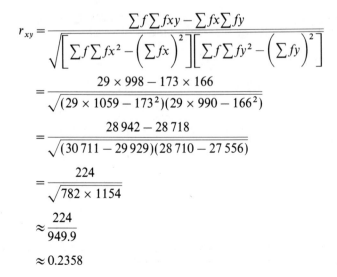

$$r_{xy} = \frac{\sum f \sum fxy - \sum fx \sum fy}{\sqrt{\left[\sum f \sum fx^2 - \left(\sum fx\right)^2\right]\left[\sum f \sum fy^2 - \left(\sum fy\right)^2\right]}}$$

$$= \frac{29 \times 998 - 173 \times 166}{\sqrt{(29 \times 1059 - 173^2)(29 \times 990 - 166^2)}}$$

$$= \frac{28\,942 - 28\,718}{\sqrt{(30\,711 - 29\,929)(28\,710 - 27\,556)}}$$

$$= \frac{224}{\sqrt{782 \times 1154}}$$

$$\approx \frac{224}{949.9}$$

$$\approx 0.2358$$

Thus, there is a low correlation between the length of the first name and the length of the surname for our group of students. Also, while we suspected a negative relationship, we have in fact found a weak positive relationship.

We can test whether this correlation coefficient is significantly different from zero by consulting tables. For a sample of size $n = 29$, there are 27 degrees of freedom. The closest value above this on the table is $n = 32$ with 30 degrees of freedom. To be significant at the 5% level, r_{xy} would have to lie outside the range ± 0.349. As the correlation coefficient lies inside this range, it is not significantly different from zero. Parents do not seem to worry about the length of their surname when choosing their child's first name.

Section 11.8 Rank Correlation

Practical 11.10 Sports Team

Ask the members of a sports team who play together regularly to rank the team members in terms of skill (giving rank number 1 to the most skilled player in the team and so on). Go along to their next match and count the number of direct passes received by each team member. This task will require two people. Rank the team members again in terms of the number of passes that they receive. Volley ball is a good game to use for this practical, as players rotate and position does not have a strong influence on the number of passes.

Practical 11.11 Photographs of Pop Stars, etc.

Collect together 10 photos of pop stars, film actors, models, etc. (either all male or all female). Ask two people to rank the photographs in order of attractiveness.

Practical 11.12 School Subjects

Obtain a list of school subjects which are *either* fourth-year subject options or sixth-form A-level choices, together with the numbers of students who chose those subjects this year. Keep totals for boys and girls separate and rank the subjects twice, once for the boys' choices and then again for the girls' choices.

Our Results for Practical 11.10

The six players in a volley ball team were ranked according to skill and the number of passes that they received during a game (Table 11.10).

Table 11.10 Rankings for a volley ball team

Player	Ranking	
	Skill	Passes
A	1	1
B	2	3
C	3	2
D	4	5
E	5	6
F	6	4

If we look at the differences d in ranks, we always find that $\sum d = 0$. However, $\sum d^2$ could be used as the basis of a measure of agreement (Table 11.11).

Table 11.11 Practical results

Ranking		d	d^2
Skill	Passes		
1	1	0	0
2	3	−1	1
3	2	1	1
4	5	−1	1
5	6	−1	1
6	4	2	4
	Total	$\Sigma d = 0$	$\Sigma d^2 = 8$

This measure d^2 will be small when there is a high level of agreement between the ranks and it will take its minimum value when there is complete agreement, as in Table 11.12.

Table 11.12 Total agreement

Ranking		d
Skill	Passes	
1	1	0
2	2	0
3	3	0
4	4	0
5	5	0
6	6	0

The minimum of $\sum d^2 = 0 + 0 + 0 + 0 + 0 + 0$. If there is a low level of agreement between the ranks, $\sum d^2$ will be large, and it will take its maximum value when there is total disagreement.

The maximum value of $\sum d^2 = (-5)^2 + (-3)^2 + (-1)^2 + 1^2 + 3^2 + 5^2$. Thus, $\sum d^2$ will discriminate between differing levels of agreement by taking values in the range from 0 to 70 (Table 11.13).

Table 11.13 Total disagreement

Ranking		d	d^2
Skill	Passes		
1	6	−5	25
2	5	−3	9
3	4	−1	1
4	3	1	1
5	2	3	9
6	1	5	25
	Total		$\Sigma d^2 = 70$

It would be useful to generate a standardised measure of agreement with a range from −1 to +1, and this will mean reversing the order as −1 has to mean complete disagreement.

We can achieve this in the following way:

Measure

complete   0 ——————————— 70 total $\sum d^2$
agreement   disagreement

(i) multiply by -1 (to reverse the *sign*)

complete   0 ——————————— -70 total $-\sum d^2$
agreement   disagreement

(ii) divide by 35 (to make the range equal to 2)

complete   0 ——————————— -2 total $\dfrac{-\sum d^2}{35}$

agreement   disagreement

(iii) add 1 (to move scale to the range from $+1$ to -1)

complete $+1$ ——————————— -1 total $1-\dfrac{\sum d^2}{35}$

agreement   disagreement

We can now calculate the rank correlation coefficient for our volley ball data:

$$1 - \frac{\sum d^2}{35} = 1 - \frac{8}{35}$$

$$\approx 1 - 0.2286$$

$$= 0.7714$$

This appears to show a high level of agreement and we shall test for significance later.

The General Formula

You may have realised that the formula which we have used so far is only valid for six pairs of ranks, but we can generate a formula for a rank correlation coefficient for any number of pairs.

The maximum value of $\sum d^2$ occurs when there is total disagreement between the two sets of ranks:

Ranking		d	d^2
Skill	Passes		
1	n	$n-1$	$(n-1)^2$
2	$n-1$	$n-3$	$(n-3)^2$
3	$n-2$	$n-5$	$(n-5)^2$
$\vdots$	$\vdots$	$\vdots$	$\vdots$
$n-2$	3	$-(n-5)$	$(n-5)^2$
$n-1$	2	$-(n-3)$	$(n-3)^2$
n	1	$-(n-1)$	$(n-1)^2$

$$\sum d^2 = (n-1)^2 + (n-3)^2 + (n-5)^2 + \ldots + (n-3)^2 + (n-1)^2$$

$$= \frac{1}{3}(n^3 - n)$$

We can use a similar method as before to achieve a standardised measure in the range from -1 to $+1$:

| complete agreement | 0 | ———— | $\frac{1}{3}(n^2 - n)$ | total disagreement | $\sum d^2$ |

(i) multiply by -1 to reverse the *sign*

| complete agreement | 0 | ———— | $-\frac{1}{3}(n^2 - n)$ | total disagreement | $-\sum d^2$ |

(ii) divide by $\frac{1}{6}(n^3 - n)$ to make the range equal to 2

| complete agreement | 0 | ———— | -2 | total disagreement | $-\dfrac{6\sum d^2}{n^3 - n}$ |

(iii) add 1 to make the range from $+1$ to -1

| complete agreement | $+1$ | ———— | -1 | total disagreement | $1 - \dfrac{6\sum d^2}{n^3 - n}$ |

The correlation coefficient for n pairs of ranks, known as Spearman's rank correlation coefficient, is

$$r_S = 1 - \frac{6\sum d^2}{n^3 - n}$$

This is often written as

$$r_S = 1 - \frac{6\sum d^2}{n(n^2 - 1)}$$

(The above proof is based on an article by Griffiths (1986); the full reference is given at the end of the chapter.)

Our Results for Practical 11.12

Table 11.14 shows the numbers of boys and girls who chose each subject in the fourth-year options at a Yorkshire comprehensive school. The table also shows the ranks for boys' and girls' choices. Rank 1 has been allocated to the most popular subjects (craft design and technology for boys and home economics for girls). There are 22 subjects; so each column contains the numbers 1–22. What we wish to do is to see whether there is any agreement between girls' and boys' choices by calculating a rank correlation coefficient.

Table 11.14 Spearman's rank correlation coefficient

Subject	Boys	Girls	Rank Boys	Rank Girls	d	d^2
Art	36	33	6	8	−2	4
Biology	33	21	8	13	−5	25
Child care	12	51	16	3	13	169
French	7	14	18	15	3	9
German	0	4	22	20	2	4
Geography	53	37	4	6	−2	4
History	37	39	5	5	0	0
Chemistry	35	29	7	9	−2	4
Physics	65	48	2	4	−2	4
Science	29	35	9	7	2	4
Textiles	6	54	19	2	17	289
Catering	5	3	20	21	−1	1
Technical drawing	25	7	11	17	−6	36
Motor vehicle technology	57	3	3	22	−19	361
Business studies	9	24	17	11	6	36
Music	1	6	21	18	3	9
Media studies	24	28	12	10	2	4
Integrated humanities	14	11	15	16	−1	1
Craft design and technology	76	20	1	14	−13	169
Home economics	27	60	10	1	9	81
Modern technology	17	5	14	19	−5	25
Information technology	22	23	13	12	1	1
					Total	1 240

$$r_S = 1 - \frac{6\sum d^2}{n(n^2 - 1)}$$

$$= 1 - \frac{6 \times 1240}{22 \times 483}$$

$$= 1 - \frac{7440}{10\,626}$$

$$\approx 1 - 0.7$$

$$= 0.3$$

Significance of Spearman's Rank Correlation Coefficient for Practical 11.10

We obtained $r_S = 0.7714$ for the rankings of six volley ball players for skill and the number of passes they received. The significance of this result may be tested using tables, which show significant values of r_S for a two-tailed test. For a sample of size $n = 6$, r_S must equal or be greater in absolute value than ± 0.886 to be significant at the 5% level. As our result is less than 0.886, we do not have a rank correlation coefficient which is significantly different from zero. Despite the fact that $r_S = 0.7714$ appears to be a high value, we have not been able to prove a significant relationship between players' rankings for skill and passes received, because our sample size is very small.

Significance of Spearman's Rank Correlation Coefficient for Practical 11.12

For Practical 11.12, we found that $r_s = 0.3$ between the rankings for choice of fourth-year option subjects for boys and girls; we find from tables that, for a sample of size $n = 22$, r_s only has to reach ± 0.423. Again, our result is not significant at the 5% level. A positive value would show some agreement between girls' and boys' choices. A negative value of r_s would show disagreement between the two sets of ranks, i.e. highly ranked items on one scale are ranked low on the other scale (and vice versa).

When to Use Which Correlation Coefficient

The product moment correlation coefficient r_{xy} assumes that there is a joint normal distribution between the values of x and y. Thus, data points should have a roughly elliptical shape on a scatter diagram.

A rank correlation coefficient does not assume this and is generally more suitable where x and y are discrete variables, or where they cannot be measured properly but may be placed in order, in terms of preference, etc.

Practical Follow-up 11.5

Practical 11.10

Investigate one or all of the following.

1. Is there any relationship between the team captain's ranking of the team members and that done by another member of the team?
2. Is there any relationship between the captain's ranking of the team members and their rank for the number of passes they receive?
3. Is there any relationship between the average rank for each person (for skill) and their rank for the number of passes they receive?

Practical 11.11

Is there any measure of agreement between the rankings of two people for your group of 10 photos?

Practical 11.12

Is there any agreement between the rankings of subject choices made by girls and boys?

Proof of the Alternative Formula for the Product Moment Correlation Coefficient

$$r_{xy} = \frac{\sum_{i=1}^{n}(x-\bar{x})(y-\bar{y})}{\sqrt{\sum_{i=1}^{n}(x-\bar{x})^2 \sum_{i=1}^{n}(y-\bar{y})^2}}$$

Since

$$\sum_{i=1}^{n}(x_i-\bar{x})^2 = \sum_{i=1}^{n}x_i^2 - \frac{\left(\sum_{i=1}^{n}x_i\right)^2}{n}$$

$$\sum_{i=1}^{n}(y_i-\bar{y})^2 = \sum_{i=1}^{n}y_i^2 - \frac{\left(\sum_{i=1}^{n}y_i\right)^2}{n}$$

and $\sum_{i=1}^{n} (x_i - \bar{x})(y_i - \bar{y})$ can be written as

$$\sum_{i=1}^{n} (x_i - \bar{x})(y_i - \bar{y}) = \sum_{i=1}^{n} x_i y_i - \sum_{i=1}^{n} \bar{x} y_i - \sum_{i=1}^{n} \bar{y} x_i + \sum_{i=1}^{n} \bar{x}\bar{y}$$

As $\bar{x}$ and $\bar{y}$ are constants,

$$\sum_{i=1}^{n} (x_i - \bar{x})(y_i - \bar{y}) = \sum_{i=1}^{n} x_i y_i - \bar{x} \sum_{i=1}^{n} y_i - \bar{y} \sum_{i=1}^{n} x_i + n\bar{x}\bar{y}$$

$$= \sum_{i=1}^{n} x_i y_i - n\bar{x}\bar{y} - n\bar{x}\bar{y} + n\bar{x}\bar{y}$$

$$= \sum_{i=1}^{n} x_i y_i - n\bar{x}\bar{y}$$

$$= \sum_{i=1}^{n} x_i y_i - \frac{\sum_{i=1}^{n} x_i \sum_{i=1}^{n} y_i}{n}$$

When these are substituted into the original formula, r_{xy} becomes

$$r_{xy} = \frac{n \sum_{i=1}^{n} x_i y_i - \sum_{i=1}^{n} x_i \sum_{i=1}^{n} y_i}{\sqrt{\left[n_i \sum_{i=1}^{n} x_i^2 - \left(\sum_{i=1}^{n} x_i \right)^2 \right]\left[n \sum_{i=1}^{n} y_i^2 - \left(\sum_{i=1}^{n} y_i \right)^2 \right]}}$$

Use of Coding: Change in Origin and Units

x and y are coded so that

$x_i = A + Bu_i$

$\bar{x} = A + B\bar{u}$

and

$y_i = C + Dv_i$

$\bar{y} = C + D\bar{v}$

Therefore,

$$r_{xy} = \frac{\sum_{i=1}^{n} (x_i - \bar{x})(y_i - \bar{y})}{\sqrt{\sum_{i=1}^{n} (x_i - \bar{x})^2 \sum_{i=1}^{n} (y_i - \bar{y})^2}}$$

$$= \frac{\sum_{i=1}^{n} (Bu_i - B\bar{u})(Dv_i - D\bar{v})}{\sqrt{\sum_{i=1}^{n} (Bu_i - B\bar{u})^2 \sum_{i=1}^{n} (Dv_i - D\bar{v})^2}}$$

$$= \frac{\sum\limits_{i=1}^{n} (u_i - \bar{u})(v_i - \bar{v})}{\sqrt{\sum\limits_{i=1}^{n} (u_i - \bar{u})^2 \sum\limits_{i=1}^{n} (v_i - \bar{v})^2}}$$

$= r_{uv}$ by definition

Range of Values Taken by r_{xy}

The limits on the value of r can be determined using Cauchy's inequality. If a_i and b_i are real numbers (for $i = 1, 2, 3, \ldots, n$). Then $\left(\sum\limits_{i} a_i b_i\right)^2 \leqslant \left(\sum\limits_{i} a_i^2\right)\left(\sum b_i^2\right)$.

Now consider the expression $\sum\limits_{i}(ka_i + b_i)^2$, where k is any real number. Since each term of the summation is squared, it will be positive. Therefore, for all k,

$$\sum\limits_{i}(ka_i + b_i)^2 \geqslant 0$$

$$\sum\limits_{i}(k^2 a_i^2 + 2ka_i b_i + b_i^2) \geqslant 0$$

$$k^2 \sum\limits_{i} a_i^2 + 2k \sum\limits_{i} a_i b_i + \sum\limits_{i} b_i^2 \geqslant 0$$

We know that for a quadratic function $ax^2 + bx + c \geqslant 0$ for all x; then $b^2 \leqslant 4ac$. If we regard $k^2 \sum\limits_{i} a_i^2 + 2k \sum\limits_{i} a_i b_i + b_i^2$ as a quadratic function in k, $\left(2 \sum\limits_{i} a_i b_i\right)^2 \leqslant 4 \sum\limits_{i} a_i^2 \sum b_i^2$.

Cancelling by 4 gives $\left(\sum\limits_{i} a_i b_i\right)^2 \leqslant \sum\limits_{i} a_i^2 \sum b_i^2$.

Now, for the correlation coefficient r,

$$r^2 = \frac{S_{xy}^2}{S_x^2 S_y^2}$$

$$= \frac{\left(\sum\limits_{i}(x_i - \bar{x})(y_i - \bar{y})\right)^2}{\sum\limits_{i}(x_i - \bar{x})^2 \sum\limits_{i}(y_i - \bar{y})^2}$$

So using Cauchy's inequality with $x_i - \bar{x} = a_i$ and $y_i - \bar{y} = b_i$, we see that $r^2 \leqslant 1$. Hence $-1 \leqslant r \leqslant 1$.

Reference

Griffiths, D., 1986, A pragmatic approach to Spearman's rank correlation coefficient, in *The Best of Teaching Statistics*, Teaching Statistics Trust, Sheffield.

Chapter 12 Linear Regression

Introduction

In Chapter 11, we examined the strength of any possible linear relationship between two variables. We now go on to find the equation of the straight line which best fits that linear relationship. We can then use this linear model for predicting future values.

We start by considering experiments in which one variable is strictly controlled and then go on to consider situations in which both variables are subject to random variation. The assumptions that we make in these two cases are quite different.

Section 12.1 Practicals

Complete as many of the practicals in this section as you can.

Practical 12.1 Distance that a Toy Car Travels

Prop up a wooden board at one end to construct a slope for a toy car to run down, or use plastic track supplied with some toy cars (the car will need to be quite heavy). Carefully measure and mark the following distances up from the bottom of the slope: 0 cm, 5 cm, 10 cm, 15 cm, 20 cm, ..., 60 cm. Place the back wheels of the car at each mark in turn, and when the car runs down the slope, measure the distance that it travels at the bottom (from the base of the slope to the back wheels of the car) (Figure 12.1). Do not push the car.

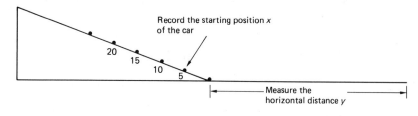

Figure 12.1   Distance that a toy car travels

For each 'run', record the starting position x up the ramp in centimetres and the horizontal distance y travelled from the base of the ramp again in centimetres. If you use the starting positions 0, 5, 10, 15, ..., 60 cm you will have 13 pairs of (x, y) values.

Practical 12.2 Springs

Hang various masses from a spring and measure the length of the spring for each mass. The masses should be used in random order (use random-number tables). Let the spring return to its natural position before putting on the next mass. We used the following masses 100 g, 200 g, 300 g, ..., 700 g.

For each trial, record the mass in grams (variable x) and the length of the spring in centimetres (variable y). Be careful not to stretch the spring beyond its elastic limit.

Practical 12.3   Lines on Paper

Draw lines 9, 12, 15, 18, 21, 24 and 27 cm long on separate sheets of paper. Give the sheets in random order to someone, asking them to draw a mark one-third of the way along the line. Measure the error for each trisection (ignoring whether the error is positive or negative). Your results should take the form of seven ordered pairs (x, y) for which x is the length of the line in centimetres and y is the absolute error.

Practical 12.4   Pack of Cards

Ask a friend to sort a pack of cards into piles as follows.

1. Place each card one at a time onto one pile on the table: 1 group.
2. Sort into red and black cards: 2 groups.
3. Sort into suits: 4 groups.
4. Split into two groups in each suit (ace to 6; 7 to king):   8 groups.
5. Split into four groups in each suit (ace, 2 and 3; 4, 5 and 6; 7, 8 and 9; 10, jack, queen and king):   16 groups.

Be sure to shuffle the pack thoroughly each time and to record the time taken for each sorting. You should have your results as five ordered pairs (x, y) where x is the number of groups and y is the time taken.

Section 12.2   Fitting a Straight Line Through the Origin

Our Results for Practical 12.1

Table 12.1 shows our results for Practical 12.1 in which a toy car was allowed to run down a ramp from various starting positions x. The horizontal distance travelled from the base of the ramp is recorded as the variable y.

Table 12.1   Results for Practical 12.1

Starting position x (cm)	0	5	10	15	20	25	30	35	40	45	50	55	60
Horizontal distance y travelled (cm)	0	1	7	16	19.5	24.5	32	34	40.5	52	53	63.5	59.5

When the results are plotted on a scatter diagram (Figure 12.2), the points lie close to a straight line, suggesting that there may be a linear relationship between x and y.

Note that the values of x, the starting positions, were pre-determined by us, while the value of y, the horizontal distances associated with each value of x, were found by experiment. For this reason, x is called the independent (or explanatory) variable and y the dependent (or response) variable.

The values of x, the starting positions of the car, have been measured accurately. However, the measurements of y, the horizontal distances travelled, are less accurate. If we had repeated the entire experiment, we would almost certainly have a different set of y values, caused by random variation. This may be due to measurement error and/or to the small changes in the environment so that for any value of x the possible y values form a

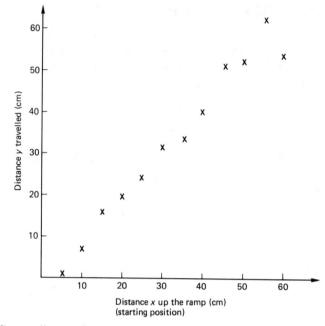

Figure 12.2 Scatter diagram for Practical 12.1

distribution. The normal distribution seems to be a suitable model for each group of y values as we would expect them to cluster around a mean value. Also extreme values will be rare and there is no reason to expect differences from the mean value to be either positive or negative.

Indeed a simpler version of this experiment is suggested in Chapter 5. If you tried Practical 5.4, you will have found that for any one starting point (usually the top of the ramp) the distances travelled by the car over about 100 trials do form an approximate normal distribution.

We assume that it is the random variation in the y values which causes the points on the scatter diagram to deviate from a straight line. If the underlying straight line were drawn on the diagram, some points would lie above the straight line and other points below it.

Figure 12.3 shows how random errors could affect two points (x_1, y_1) and (x_2, y_2).

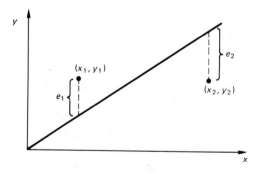

Figure 12.3 How random errors could affect two points (x_1, y_1) and (x_2, y_2)

The point (x_1, y_1) lies above the straight line by a difference of e_1 while (x_2, y_2) lies below the straight line by a difference of e_2. One will be a positive error, and the other negative.

If the car is placed at the bottom of the ramp so that $x = 0$, y will also be 0. So, for this experiment, we know that the possible straight-line relationship between x and y must pass through $(0, 0)$. Thus the straight line must take the form $y = \beta x$. As it can be used to predict y values from x, it is known as the regression line of y on x. Each point on our scatter diagram (x_i, y_i) can be expressed as $y_i = x_i + e_i$, where x_i is the value of y which is predicted by the straight-line equation and e_i is a random error. We shall assume that the e_i values are random errors independent of x_i and normally distributed with a mean 0.

We now consider how to find the equation of the line

$$y = \beta x$$

The principal that we used to estimate β is to find the line that minimises $\sum_i e_i^2$. This is known as the method of least squares. Take any line $y = bx$ and express $\sum_i e_i^2$ as a function of b and find the value of b that minimises this expression. This value of b is our estimate of the unknown β.

For the line $y = bx$,

$$y_i = bx_i + e_i$$

$$e_i = y_i - bx_i$$

$$\sum_i e_i^2 = \sum_i (y_i - bx_i)^2$$

We wish to differentiate with respect to b as it is b which can be changed to minimise $\sum_i e_i^2$:

$$\frac{d\left(\sum_i e_i^2\right)}{db} = \sum_i [-2x_i(y_i - bx_i)]$$

$$0 = \sum_i x_i(y_i - bx_i)$$

$$0 = \sum_i (x_i y_i - bx_i^2)$$

$$0 = \sum_i x_i y_i - \sum_i bx_i^2$$

$$b\sum_i x_i^2 = \sum_i x_i y_i$$

$$b = \frac{\sum_i x_i y_i}{\sum_i x_i^2}$$

Thus the least-squares regression line of y on x passing through $(0, 0)$ is of the form

$$y = bx$$

where

$$b = \frac{\sum_i x_i y_i}{\sum_i x_i^2}$$

A Worked Example

We can calculate the regression line of y on x for our experimental data from Practical 12.1, assuming that it passes through the origin (Table 12.2).

Table 12.2   Calculation regression line for Practical 12.1 data

Starting position x (cm)	Distance y travelled (cm)	x^2	xy
0	0	0	0
5	1	25	5
10	7	100	70
15	16	225	240
20	19.5	400	390
25	24.5	625	612.5
30	32	900	960
35	34	1 225	1 190
40	40.5	1 600	1 620
45	52	2 025	2 340
50	53	2 500	2 650
55	63.5	3 025	3 492.5
60	59.5	3 600	3 570
390	402.5	16 250	17 140

Now

$$b = \frac{\sum_i x_i y_i}{\sum_i x_i^2}$$

$$= \frac{17\,140}{16\,250}$$

$$\approx 1.055$$

Thus, our least-squares regression line is

$$y = 1.055x$$

This line is drawn on the scatter diagram shown in Figure 12.4.

As the points on the diagram lie very close to our regression line, we can use the relationship $y = 1.055x$ to predict the horizontal distances which the car would travel from different starting points. For instance, if we placed the car at a starting point 12 cm up the ramp, we would predict that it would travel a horizontal distance of

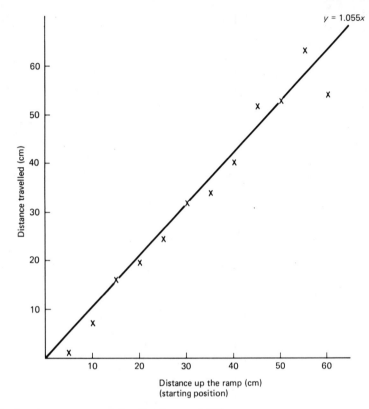

Figure 12.4   Least-squares regression line for $y = 1.055x$

$y = 1.055 \times 12$ cm

$= 12.66$ cm

In any situation such as this, it is unwise to make predictions far outside the range of the original set of measurements since there the underlying model may cease to be linear. It is not usual to force the regression line to pass through the origin even when, as in the above case, there seem to be logical reasons for doing so. Take this as an introductory example to illustrate the basic ideas behind fitting a regression line by the least-squares principle.

Section 12.3   The General Model for Fitting a Straight Line

Our Results for Practical 12.2

Table 12.3 gives our results for Practical 12.2 in which the length of a spring was measured as different masses were suspended from it.

Table 12.3   Results for Practical 12.2

Mass x (g)	100	200	300	400	500	600	700
Length y (cm)	22.9	27.3	32.2	36.5	42.0	47.0	53.2

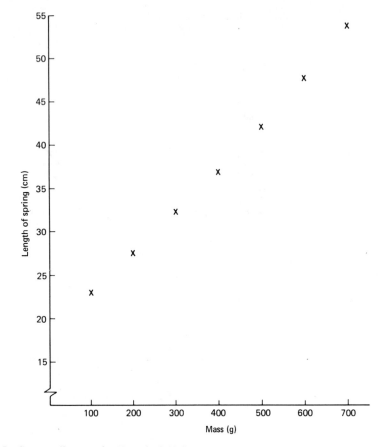

Figure 12.5   Scatter diagram for Practical 12.2

When the points are plotted on a scatter diagram (Figure 12.5), they appear to lie very close to a straight line.

The mass x hung from the spring is the independent variable which takes pre-selected values accurately known, while the length y of the spring is the dependent variable subject to random errors.

As the line does not pass through the origin, we assume that the underlying relationship is of the form

$$y = \alpha + \beta x$$

where α is the intercept on the y axis and β is the gradient of the line.

Each point (x_i, y_i) may be expressed as its predicted value calculated from the straight-line equation plus a random error. Thus,

$$y_i = \alpha + \beta x_i + e_i$$

We make the same assumptions about the distributions of y_i and e_i as outlined in Section 12.2. Again we do not know the values of α and β; so we estimate them by finding the line $y = ax + b$ which minimises $\sum_i e_i^2$ from it. We use these values of a and b as estimates of α and β.

The General Least-squares Model

For any line

$$y = ax + b$$

$$y_i = a + bx_i + e_i$$

so that

$$e_i = y_i - a - bx_i$$

and

$$\sum_i e_i^2 = \sum_i (y_i - a - bx_i)^2$$

We have two quantities a and b which can be varied to minimise $\sum_i e_i^2$ and we shall partially differentiate with respect to each separately. First, differentiate with respect to a:

$$\frac{\partial \left(\sum_i e_i^2 \right)}{\partial a} = \sum_i [-2(y_i - a - bx_i)]$$

This will equal 0 at a minimum. Then, differentiate with respect to b:

$$\frac{\partial \left(\sum_i e_i^2 \right)}{\partial a} = \sum_i [-2x(y_i - a - bx_i)]$$

Again this equals 0 at a minimum.

We now have two simultaneous equations:

$$\sum_i [-2(y_i - a - bx_i)] = 0$$

and

$$\sum_i [-2x(y_i - a - bx_i)] = 0$$

These can be written as

$$\sum_i y_i - na - b\sum_i x_i = 0$$

and

$$\sum_i x_i y_i - \sum_i ax_i - b\sum_i x_i^2 = 0$$

or, more usually,

$$\sum_i y_i = na + b_i \sum_i x_i$$

$$\sum_i x_i y_i = ax_i + b\sum_i x_i^2$$

These are known as the *normal equations* and, by solving them as simultaneous equations, we can find a and b to give the regression line $y = a + bx$ which is our estimate of the true regression line $y = \alpha + \beta x$.

Solving the normal equations simultaneously is one method of finding the equation of the regression line. Alternatively, b may be calculated directly:

$$b = \frac{n\sum_i x_i - \sum_i x_i \sum_i y_i}{n\sum_i x_i - \left(\sum_i x_i\right)^2}$$

This formula comes from solving the normal equations as simultaneous equations. To find a, look at the first normal equation

$$\sum_i y_i = na + b\sum_i x_i$$

Divide throughout by n to give

$$\bar{y} = a + b\bar{x}$$

This shows that the least-squares regression line always passes through $(\bar{x}, \bar{y})$ and we may also substitute in to find the value of a. (An alternative proof which does not use partial differentiation is shown at the end of this chapter.)

A Worked Example

We use the data from Practical 12.2 to give Table 12.4.

Table 12.4   Data from Practical 12.2

Mass x	Length y	xy	y^2
100	22.9	2 290	10 000
200	27.3	5 460	40 000
300	32.2	9 660	90 000
400	36.5	14 600	160 000
500	42.0	21 000	250 000
600	47.0	28 200	360 000
700	53.2	37 240	490 000
2800	261.1	118 450	1 400 000

The normal equations are

$$\sum_i y_i = na + b\sum_i x_i \tag{12.1}$$

$$\sum_i x_i y_i = a\sum_i x_i + b\sum_i x_i^2 \tag{12.2}$$

Substituting in values gives

$$261.1 = 7a + 2800b \tag{12.1a}$$

$$118\,450 = 2800a + 1\,400\,000b \tag{12.2a}$$

To eliminate b, multiply equation (12.1a) by 500:

$$130\,550 = 3500a + 1\,400\,000b \qquad\qquad (12.1b)$$

If we subtract equation (12.2a) from equation (12.1b), we obtain

$$17.29 = a$$

To find b, substitute for a in equation (12.1a):

$$261.1 = 17.29 \times 7 + 2800b$$

$$261.1 = 121.03 + 2800b$$

$$140.07 = 2800b$$

$$0.05 = b$$

This gives us the equation of our regression line as

$$y = 17.29 + 0.05x$$

This line may be used to predict value of y (i.e. the length of the spring) for given x values (masses). To calculate the predicted length of the spring when 550 g is suspended from it, substitute $x = 550$:

$$y = 17.29 + 0.5 \times 550$$

$$y = 17.29 + 27.5$$

$$y = 44.79$$

Thus the predicted length of the spring would be 44.79 cm.

Even if we had been measuring the extension in length of the spring rather than the total length as in this example, we would not have forced the regression line to pass through the origin. In fact, with this type of practical the relationship between mass and spring extension is not strictly linear around zero.

Section 12.4   A Variation of the Linear Model

Our Results for Practical 12.4

For this practical a subject was asked to sort a pack of cards into groups and the time taken was recorded. Table 12.5 shows our results.

Table 12.5   Results for Practical 12.4

Number x of groups	1	2	4	8	16
Time taken (s)	27	42	55	100	125

The scatter diagram (Figure 12.6) drawn for these results seems to show a curvilinear rather than a straight-line relationship.

This practical is used by psychologists to illustrate Hicks' law which states the relationship between the complexity of a task and the time taken to complete it. According to Hicks' law, the time taken for a subject to sort cards into 4 groups should be about twice the time

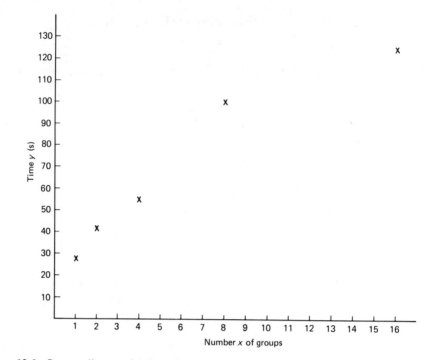

Figure 12.6   Scatter diagram for Practical 12.4

taken to sort cards into 2 groups. However, it should only take them about three times as long to sort cards into 8 groups and four times as long for 16 groups.

This can be expressed more simply in mathematical terms. Let

t be the time taken for 2 groups

Then

$2t \propto$ time taken for 4 groups (note that $4 = 2^2$)

$3t \propto$ time taken for 8 groups (note that $8 = 2^3$)

$4t \propto$ time taken for 16 groups (note that $16 = 2^4$)

Thus the time taken should be in proportion to the number of groups expressed as a power of 2. If we adjust the x axis on our scatter diagram to a logarithmic scale in base 2, a straight line begins to emerge (Table 12.6 and Figure 12.7).

Table 12.6   Data for Practical 12.4 on a logarithmic scale

Number x of groups	$W = \log_2 x$	Time y (s)
1	0	27
2	1	42
4	2	55
8	3	100
16	4	125

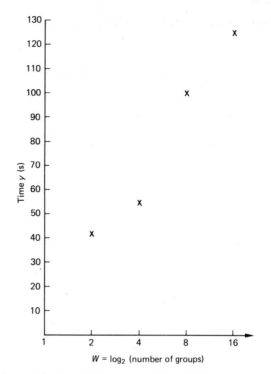

Figure 12.7 Scatter diagram for Practical 12.4 on a logarithmic scale

Calculating a Regression Line

We can use the linear regression model provided that we use $\log_2 x\,(=W)$ as the independent variable rather than x itself. We can conclude the regression line of y on W in this case (Table 12.7).

Table 12.7 Calculation of regression line for Practical 12.4 data on a logarithmic scale

Number x of groups	$W = \log_2 x$	Time y (s)	W^2	Wy
1	0	27	0	0
2	1	42	1	42
4	2	55	4	110
8	3	100	9	300
16	4	125	16	500
Total	10	349	30	952

The normal equations are

$$\sum y = na + b\sum W \tag{12.3}$$

$$\sum Wy = a\sum W + b\sum W^2 \tag{12.4}$$

Substituting values into equations (12.3) and (12.4) gives

$$349 = 5a + 10b \tag{12.3a}$$

$$952 = 10a + 30b \tag{12.4a}$$

Multiply equation (12.3a) by 2 to give

$$698 = 10a + 20b \qquad (12.3b)$$

If we subtract equation (12.3b) from equation (12.4a), we obtain

$$254 = 10b$$

$$b = 25.4$$

To find a, substitute for b in equation (12.3a):

$$349 = 5a + 254$$

$$95 = 5a$$

$$19 = a$$

Thus the regression line for y on W is

$$y = 19 + 25.4W$$

As W is $\log_2 x$, we have

$$y = 19 + 25.4 \log_2 x$$

This line can be used to predict the time taken for sorting cards into 32 groups.
As $\log_2 32 = 5$, we therefore substitute $W = 5$ to give

$$y = 19 + 25.4 \times 5$$

$$y = 19 + 127$$

$$y = 146$$

Thus the predicted time is 146 s.

Practical Follow-up 12.1

1. Calculate the regression line to predict y from x for each of the practical experiments for which you have collected data.
2. Draw a scatter diagram (if you have not already done so) and draw your regression line on it. Remember that the line will pass through $(\bar{x}, \bar{y})$.

Section 12.5   Bivariate Distributions

In all the practicals outlined in Section 12.1, one variable (usually y) was subject to random variation while the other variable x was controlled. Situations such as these usually occur only in some type of scientific experiment.

There are many more situations in which both variables x and y are subject to random variation. Neither variable can be strictly controlled; so neither x nor y can be treated as the independent variable in the sense that we have used so far.

The practicals in this section all yield data of this second type. If you have already worked through Chapter 11, you may have completed one or more of these practicals and therefore already have data available for the regression analysis in this section.

Practical 12.5   Lengths and Widths of Leaves

For a random sample of 15 leaves (of the same species), measure the length and width of each leaf. You may have these data already available from your practical work for Chapter 5.

Practical 12.6   Heights and Weights of Students

Measure and record *either* of the following.

1. Heights and weights of 15 students of the same sex.
2. The heights and ages for a sample of younger children of various ages.

Practical 12.7   Heights of Students and their Parents

Measure the heights of 15 students of the same sex and find also *either* of the following.

1. The heights of their fathers.
2. The heights of their mothers.

Practical 12.8   School Tuck-shop

Conduct a survey at your school tuck-shop. For each customer, note down the following.

1. The amount of money that they spend.
2. Their age (or year group in the school).
3. Ask them the amount of pocket money that they receive each week.

   If you collect up a large number of results, you may need to take a random sample of about 20 of all your replies.

Practical 12.9   Weights of Coins

Weigh 15 coins of the same denomination (2 p, 10 p or 50 p are suggested) and record their weights and ages. An electronic balance accurate to within 0.01 g is needed for this practical.

Practical 12.10   Daily Rainfall and Sunshine Data

Obtain data for daily rainfall and hours of sunshine from your local weather station for a period of at least 2 weeks.

Our Results for Practical 12.5

Table 12.8 shows the lengths and widths of 10 leaves.

Table 12.8   Lengths and widths of leaves

Length x (mm)	103	119	146	148	87	134	151	147	122	129
Width y (mm)	35	37	41	53	35	39	50	42	33	43

   In Figure 12.8, we have two variables x and y which are correlated ($r_{xy} = 0.748$) but neither variable has been controlled experimentally and neither x nor y can strictly be called the independent variable. Both x and y are subject to random errors in measurement and random variations. However, we can use a regression model provided that we are able

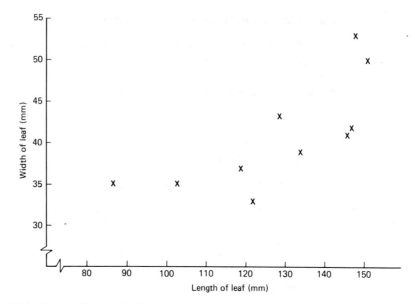

Figure 12.8 Scatter diagram for Practical 12.5

to assume that x and y are jointly normally distributed. Each of the variables (length and width) must be normally distributed and for each length of leaf there must be a normal distribution of widths (and vice versa). This is quite a difficult idea to visualise but Figure 12.9 may help.

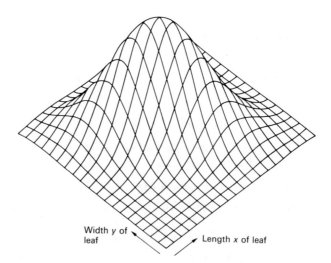

Figure 12.9 Joint normal distribution of the lengths and widths of leaves

The regression line of y on x (to predict width from length) will pass through the mean width for each chosen value of x (length) and, from this line, we can predict the *mean* width for any given length. The regression line should therefore be known as the regression line for $\bar{Y}|X = x$.

In practical terms, we tend to assume that X and Y are jointly normally distributed provided that both are variables with continuous measures which could separately be assumed to be normal and provided also that the points on the scatter diagram lie within an approximate elliptical shape.

Predicting y from x

The regression model for $\bar{y}$ on x takes the form

$$\bar{y} = \alpha + \beta x$$

Owing to random variations, sampled points will not lie exactly on the regression line and each point (x_i, y_i) can be expressed as

$$y_i = \alpha + \beta x_i + e_i$$

where e_i is the distance of the point (x_i, y_i) above or below the line. Since we do not know the true values of α and β for the regression line, we estimate them by finding the line which minimises the sum of squares $\sum_i e_i^2$. This will give us a line

$$y = a + bx$$

and the resulting normal equations are

$$\sum_i y_i = na + b \sum_i x_i$$

and

$$\sum_i x_i y_i = a \sum_i x_i + b \sum_i x_i^2$$

These give the slope of the line as

$$b = \frac{n \sum_i x_i y_i - \sum_i x_i \sum_i y_i}{n \sum_i x_i^2 - \left(\sum_i x_i\right)^2}$$

and a can be found from $\bar{y} = a + b\bar{x}$. We are using a and b to estimate the unknown α and β.

Predicting x from y

It is also possible to calculate a line to predict an x value for a given y value, although in fact we find the mean value of x for any value of y. Here we require a linear model of the form (Figure 12.10)

$$\bar{X} = \alpha' + \beta' y$$

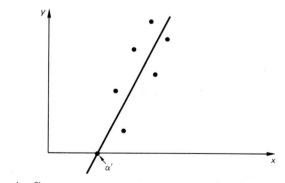

Figure 12.10 $\bar{X} = \alpha' + \beta' y$

Strictly speaking, this line should be known as the equation for $\bar{X}\,|\,Y = y$ rather than the regression line of x on y.

Note that α' is the intercept on the x axis and β' is $1/$gradient of the line (i.e. the gradient measured as increase in x divided by increase in y).

We know that the points on the scatter diagram will not lie exactly on the line, but in this case we wish to minimise the errors made in predicting values of x. These errors are shown by horizontal differences from the regression line (Figure 12.11).

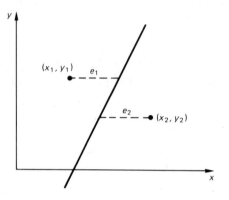

Figure 12.11 Errors on the scatter diagram in Figure 12.10

Thus each observed point (x_i, y_i) may be written as

$$x_i = \alpha' + \beta' y_i + e_i$$

By following the same method of least squares shown in Section 12.3, we can minimise $\sum_i e_i^2$ and arrive at the normal equations for regression of x on y, using a' and b' to estimate α' and β'

$$\sum_i x_i = na' + b' \sum_i y_i$$

$$\sum_i x_i y_i = a' \sum_i y_i + b' \sum_i y_i^2$$

These can be solved simultaneously to find a' and b'. Alternatively,

$$b' = \frac{n \sum_i x_i y_i - \sum_i x_i \sum_i y_i}{n \sum_i y_i^2 - \left(\sum_i y_i\right)^2}$$

and the regression line again passes through $(\bar{x}, \bar{y})$, so that a' may be found by substitution.

A Worked Example for Practical 12.6 (Working with a Change in Units)

To make working easier, the data can be coded so that

$$u = x - 100$$

$$v = y - 40$$

as in Table 12.9.

Table 12.9 Data from Practical 12.6 after a change in units

Length x (mm)	Width y (mm)	$u = x - 100$	$v = y - 40$	u^2	v^2	uv
103	35	3	-5	9	25	-15
119	37	19	-3	361	9	-57
146	41	46	1	2 116	1	46
148	53	48	13	2 304	169	624
87	35	-13	-5	169	25	65
134	39	34	-1	1 156	1	-34
151	50	51	10	2 601	100	510
147	42	47	2	2 209	4	94
122	33	22	-7	484	49	-154
129	43	29	3	841	9	87
1 286	408	286	8	12 250	392	1 166

To predict the width from the length, we calculate the regression line of v on u and substitute back to find the regression line of y on x.

Regression Line of v on u

The normal equations are

$$\sum v = na + b\sum u \tag{12.5}$$

$$\sum uv = a\sum u + b\sum u^2 \tag{12.6}$$

Substituting in values gives

$$8 = 10a + 286b \tag{12.5a}$$

$$1166 = 286a + 122\,50b \tag{12.6a}$$

Multiply equation (12.5a) by 28.6:

$$228.8 = 286a + 8179.6b \tag{12.5b}$$

If we subtract equation (12.5b) from equation (12.6a), we obtain

$$937.2 = 4070.4b$$

$$0.23 = b$$

To find a, substitute for b in equation (12.5a):

$$8 = 10a + 286 \times 0.23$$

$$8 = 10a + 65.78$$

$$-57.78 = 10a$$

$$-5.778 = a$$

Thus,

$$v = -5.778 + 0.23u$$

Regression line of y on x

When $v = y - 40$ and $u = x - 100$ is substituted in $v = -5.778 + 0.23u$, we obtain

$$y - 40 = -5.778 + 0.23(x - 100)$$

$$y - 40 = -5.778 + 0.23x - 23$$

$$y = 11.22 + 0.23x$$

To predict the length from the width, we require the regression line of x on y. First, we calculate the regression line of u on v and then substitute into $u = 26.66 + 2.43v$.

Regression line of x on y

If we substitute $u = x - 100$ and $v = y - 40$ in $u = 26.66 + 2.43v$, we obtain

$$x = 29.46 + 2.43y$$

This line may be used to predict the length of a leaf which is 50 mm wide, say, by substituting in $y = 50$:

$$x = 29.46 + 2.43 \times 50$$

$$x = 29.46 + 121.5$$

$$x = 150.96$$

Thus, we would predict a length of 150.96 mm.

Practical Follow-up 12.2

Work through the following steps for each set of bivariate data you have collected.

1. Draw a scatter diagram.
2. Calculate r_{xy}. What does this tell you about the predictive value of your regression lines?
3. Calculate the regression lines of y on x and x on y. Draw them on your scatter diagram remembering that *both* lines will pass through $(\bar{x}, \bar{y})$.

Algebraic Method for Least Squares

$$\sum_i e_i^2 = \sum_i (y_i - \alpha - \beta x_i)^2$$

The algebra for this least-squares method is simplified if coding is used so that

$$u_i = x_i - \bar{x}$$

and

$$v_i = y_i - \bar{y}$$

Note that $\sum_i u_i = 0$ and $\sum_i v_i = 0$. Now

$$e_i = y_i - \alpha - \beta x_i,$$

$$= (y_i - \bar{y}) - \beta(x_i - \bar{x}) - (\alpha - \bar{y} + \beta x)$$

$$= v_i - \beta u_i - (\alpha - \bar{y} + \beta \bar{x})$$

This gives us

$$\sum_i e_i^2 = \sum_i (y_i - \alpha - \beta x_i)^2$$

$$= \sum_i [v_i - \beta u_i - (\alpha - \bar{y} + \beta \bar{x})]^2$$

The cross-product term in the expansion of this expression is

$$-2\sum_i (v_i - \beta u_i)(\alpha - \bar{y} + \beta \bar{x}) = -2(\alpha - \bar{y} + \beta \bar{x})\sum_i (v_i - \beta u_i)$$

as $\alpha - \bar{y} + \beta \bar{x}$ is a constant and can be brought in front of the $\sum$.

$$-2(\alpha - \bar{y} - \beta \bar{x})\sum_i (v_i - \beta u_i) = -2(\alpha - \bar{y} + \beta \bar{x})\left(\sum_i v_i - \beta \sum_i u_i\right)$$

$$= 0$$

As $\sum_i v_i = 0$ and $\sum_i u_i = 0$, the cross-product therefore equals 0

$$\sum_i e_i^2 = \sum_i (v_i - \beta u_i)^2 + \sum (\alpha - \bar{y} + \beta \bar{x})^2$$

$$= \sum_i (v_i - \beta u_i)^2 + n(\alpha - \bar{y} + \beta \bar{x})^2$$

Therefore, $\sum_i e_i^2$ will be a minimum when each of those two terms is a minimum.

The first term is $\sum_i (v_i - \beta u_i)^2$. For this to be a minimum, we can differentiate with respect to β (as in Section 12.2) and we find

$$\beta = \frac{\sum_i u_i v_i}{\sum_i u_i}$$

$$\beta = \frac{\sum_i (x_i - \bar{x})(y_i - \bar{y})}{\sum_i (x_i - \bar{x})^2}$$

The second term is $n(\alpha - \bar{y} + \beta \bar{x})^2$. This will take the minimum value of zero if $\alpha = \bar{y} - \beta \bar{x}$.

Chapter 13 χ^2: Are Things as (We Think) They Should Be?

Introduction

In this chapter, we develop a test to determine whether data gathered for an investigation or project are close to those which would be generated by a mathematical model. The model will be one which we have chosen, guided by a theory or hypothesis and which we feel should be suitable for explaining the data and making future predictions.

When you have finished working through this chapter, you should be able to achieve the following.

1. Test observed data against a theoretical distribution, and be able to answer questions such as 'Can the heights of female students be assumed to come from a normal distribution?'.
2. Compare the similarity of two distributions.

Section 13.1 Practicals

Complete all three practicals.

Practical 13.1   **Random Numbers**

Problem: **Can you write down random numbers?**

Quickly write down 100 single-digit numbers at random—using the figures 0–9. You may have done this practical already, if you have worked through Chapter 6. (Do not use random-number tables or a calculator or microcomputer.)

Form a frequency table for the number of times that you have written each figure.

Practical 13.2   **Random-number Tables**

Problem: **Are random-number tables random enough?**

Find a table of random numbers or use the random-number generator on a calculator or microcomputer. Write down 100 random numbers from one of these sources and again form a frequency table for the number of times that each figure appears. Each student in the class should use a different set of random numbers.

Practical 13.3   **'Favourite' Numbers**

Problem: **Are 'favourite' numbers random?**

Ask 100 people to tell you their 'favourite' number. Make a frequency table for those numbers up to 10, but ignore numbers over 10.

Section 13.2 Our Results

Our results for Practical 13.2 are shown in Table 13.1.

Table 13.1 Results for Practical 13.2

2	2	7	7	8	8	3	3	1	7	7	8	0	8	9	2	7	3	4	9
0	3	6	4	5	9	0	7	4	2	9	5	8	1	3	9	0	6	4	1
2	0	8	1	9	2	3	4	5	1	9	0	3	5	0	8	2	1	4	2
6	2	4	9	0	0	9	0	6	7	8	6	9	3	4	8	3	1	8	2
1	9	0	7	6	7	6	8	4	9	0	3	2	7	4	7	5	2	0	3

These figures are 100 numbers taken from a random-number table. Each figure is supposed to be an independent random sample from a population in which the numbers 0–9 are equally likely, so that each has a probability of 1/10 of appearing at each selection. So we expect to have approximately 10 of each figure. For example,

expected frequency for $3 = p(3) \times$ total frequency

$$= \frac{1}{10} \times 100$$

$$= 10$$

However, this expected frequency should be taken only as a rough guide, and over a sample of only 100 random numbers there will be some discrepancies. If there was a constraint so that there *had* to be exactly 10 of each figure, the numbers would no longer be random. (Should we find that we have a fairly even distribution of numbers, we cannot definitely conclude that we do have genuine random numbers from this test alone. For example the numbers 01, 02, 03, 04, 05, 06, 07, 08, 09, 10, 11, ..., 97, 98, 99 satisfy this test but are far from random.)

Random numbers have other important properties—refer to Chapter 6. Table 13.2 shows the actual frequencies with which each figure appeared in our sample of random numbers.

Table 13.2 Actual frequencies for our random numbers

Number	Frequency
0	13
1	8
2	12
3	11
4	10
5	5
6	7
7	11
8	11
9	12
Total	100

In Figure 13.1, we have shown the actual (observed) frequencies together with the expected frequency (10) for each figure.

If we wish to test whether the distribution of this sample of 100 random numbers is radically different from the expected distribution, we can calculate the differences between the observed and expected frequencies (Table 13.3).

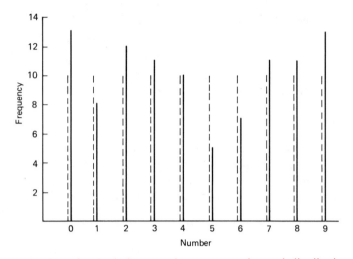

Figure 13.1   Distribution of 100 random numbers: ———, observed distribution; ---, expected distribution

Table 13.3   Differences between the observed and expected frequencies

Number	Observed frequency O	Expected frequency E	Difference $O - E$
0	13	10	3
1	8	10	-2
2	12	10	2
3	11	10	1
4	10	10	0
5	5	10	-3
6	7	10	-3
7	11	10	1
8	11	10	1
9	12	10	2
		Total	0

The total of the differences $O - E$ is zero as the positive differences and negative differences cancel each other out, and this will always be the case.

To overcome this problem the differences, $O - E$ may be squared. In this example, each number had an equal expected frequency, but this will not always be so. If we are to construct a test using the total of these square differences, we shall wish to assign a greater importance to the same difference from a small expected frequency and less importance to that same difference from a higher expected frequency (i.e. a difference of 5 must be more significant if the expected frequency is 10 than if it is 100).

One way of allowing for this is to divide each squared difference by the expected frequency for that category.

The test statistic that we use is

$$\sum \frac{(O - E)^2}{E}$$

The larger the difference between the observed and expected values, then the larger will be the value of $\sum[(O-E)^2/E]$.

For our set of frequencies,

$$\sum\frac{(O-E)^2}{E}=\frac{3^2}{10}+\frac{(-2)^2}{10}+\frac{2^2}{10}+\frac{1^2}{10}+\frac{0^2}{10}+\frac{(-5)^2}{10}+\frac{(-3)^2}{10}+\frac{1^2}{10}+\frac{1^2}{10}+\frac{2^2}{10}$$

$$=0.9+0.4+0.4+0.1+0+2.5+0.9+0.1+0.1+0.4$$

$$=5.8$$

Clearly, different sets of 100 random numbers will usually give different values of $\sum[(O-E)^2/E]$. We could find the distribution of all the values of this test statistic based on all the different sets of 100 random numbers possible. But this is not necessary since the sampling distribution of $\sum[(O-E)^2/E]$ is very close to the theoretical distribution χ^2 (or chi squared). In fact, there is a family of χ^2 distributions each with a different shape depending on the number of 'degrees of freedom' denoted by v (pronounced 'nu'). For any χ^2 distribution, the number of degrees of freedom indicates the number of independent free choices in allocating the expected frequencies. In our example there are 10 expected frequencies (one for each of the figures 0–9). However, only nine can vary independently; the tenth is determined by the fact that the total frequency must equal 100. The fact that our expected frequencies must come to the same total as our observed frequencies is known as a 'constraint'.

To calculate the number of degrees of freedom,

v = number of classes or groups − number of constraints

Here we have 10 classes and 1 constraint; so $v=9$.

The shape of the χ^2 distribution is different for each value of v and the distribution function is very complicated (Figure 13.2). The mean of χ_v^2 is v, and the variance is $2v$. The distribution is generally positively skewed so that the mean is to the right of the mode. For large values of v the distribution becomes approximately symmetrical (see Figure 13.2).

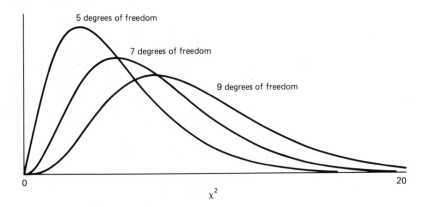

Figure 13.2   The χ^2 distribution for various values of v

Section **13.3   Significance Testing Using the χ^2 Distribution**

For our sample of random numbers the value of χ^2 was 5.8. We might like to see where this value of χ^2 comes in the χ^2 distribution with 9 degrees of freedom. By comparing this sample value of χ^2 with the distribution of χ^2, we can decide whether our expected frequencies are significantly different from our observed frequencies. In other words, we use χ^2 as our test statistic to test the *goodness of fit* between the two. Clearly for larger values of χ^2 the sample is less likely to have come from the theoretical distribution.

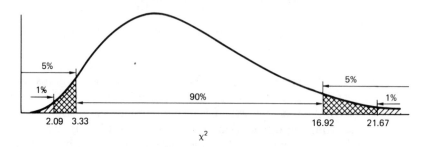

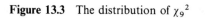

Figure 13.3   The distribution of $\chi_9{}^2$

We can find specific areas under the curve from tables. For $\chi_9{}^2$ (Figure 13.3),

$P(\chi^2 > 2.09) = 0.99$ (1% of the distribution lies to the *left* of 2.09)

$P(\chi^2 > 3.33) = 0.95$ (5% of the distribution lies to the *left* of 3.33)

The left-hand end of the distribution is rarely used for hypothesis testing. A very low value for χ^2 implies that the observed values are extremely close to the expected values, so close in fact that we should be suspicious. Such 'perfect' data may well be fictitious and have a very low probability of occurring as a random sample.

As a high value for χ^2 implies a poor fit between the observed and expected frequencies, it is the right-hand end of the distribution which is used for most hypothesis testing.

For $\chi_9{}^2$,

$P(\chi^2 > 16.92) = 0.05$

$P(\chi^2 > 21.67) = 0.01$

Our null hypothesis is that there is no difference between the observed frequencies from our sample and the theoretical frequencies.

Working with 9 degrees of freedom, we require a χ^2 total of 16.92 to find a significant difference at the 5% level. This means that we shall make a type 1 error (i.e. reject H_0 when it is true) for 5% of all possible samples of genuine random numbers. However, as this probability is very low, we are more likely to infer that the numbers in the sample were not true random numbers after all. As we do not have a clear alternative hypothesis, we cannot work out the probability of a type 2 error. We cannot find the probability of accepting a 'fictitious' set of numbers as genuine random numbers.

The sample of 100 random numbers, which we took from a random-number table, generated a χ^2 total of 5.8, when their frequencies were compared with their theoretical expected frequencies. This value for χ^2 is well within the centre of the distribution and does not fall in either of the critical regions at each end of the distribution which is not significantly different from the expected frequency distribution for random numbers. We shall accept them as being true random numbers as far as this test is concerned.

Practical Follow-up 13.1

Find the frequency table for the 100 random numbers that you obtained from statistical tables or from a random-number generator on a calculator or microcomputer (Practical

13.2). Compare your observed frequencies with the theoretical expected frequencies and calculate the total for χ^2 where

$$\chi^2 = \sum \frac{(O-E)^2}{E}$$

Is your result significant at the 5% level? Remember that, if 100 students were to conduct this testing procedure, we would expect 5 of them to obtain results which are significant, by virtue of the high values of χ^2. It is quite possible then that a student in the class has obtained a significant result using a set of genuine random numbers.

Section 13.4 Our Results for Practical 13.1

Table 13.4 shows 100 figures which were written down by a person trying to generate random numbers of their own. (This set of figures appears in Chapter 6.)

Table 13.4 Set of 100 figures

6	4	2	0	9	8	1	3	5	2.
7	3	4	2	5	9	0	2	6	5
8	9	6	7	4	0	3	2	1	5
5	6	3	0	2	1	9	5	3	0
4	2	0	9	8	6	3	7	0	1
5	0	3	4	9	8	3	6	2	4
8	6	2	1	0	3	5	8	6	3
9	0	5	6	3	5	0	1	3	2
2	5	8	6	9	3	2	0	6	5
5	3	0	2	6	1	9	8	6	3

Table 13.5 is the frequency table showing the distribution of each of the figures 0–9.

H_0: these are true random numbers, $p = 1/10$

H_1: these are not true random numbers, $p \neq 1/10$

Testing at the 5% level the critical value of χ^2 for $v = 9$ degrees of freedom is 16.92.

Table 13.5 Frequency table of the distribution of the figures 1–9

Figure	Frequency
0	13
1	7
2	13
3	15
4	6
5	13
6	13
7	3
8	8
9	9
Total	100

Table 13.6 Differences between the observed and the expected frequencies

Number	Observed frequency O	Expected frequency E	$O - E$
0	13	10	3
1	7	10	-3
2	13	10	3
3	15	10	5
4	6	10	-4
5	13	10	3
6	13	10	3
7	3	10	-7
8	8	10	-2
9	9	10	-1

The differences $O - E$ are given in Table 13.6.

$$\chi^2 = \sum \frac{(O - E)^2}{E}$$

$$= \frac{3^2}{10} + \frac{(-3)^2}{10} + \frac{3^2}{10} + \frac{5^2}{10} + \frac{(-4)^2}{10} + \frac{3^2}{10} + \frac{3^2}{10} + \frac{(-7)^2}{10} + \frac{(-2)^2}{10} + \frac{(-1)^2}{10}$$

$$= 0.9 + 0.9 + 0.9 + 2.5 + 1.6 + 0.9 + 0.9 + 4.9 + 0.4 + 0.1$$

$$= 14$$

This value of χ^2 is not significant at the 5% level, although it is high. We would do well to be suspicious although, on the *basis of this test*, we cannot reject them as random numbers (we must accept H_0 under the criteria that we have set up). We might well wish to obtain a larger sample to see what happens then. We could also develop other tests to determine whether a set of numbers are truly random. (Refer back to Chapter 6.)

Note

The χ^2 test should only be used for tables of *frequencies* and *not* for tables of measurements such as lengths, weights, etc., where the magnitude of the numbers will depend on the units used. Any result for χ^2 would be invalid in that case.

Practical Follow-up **13.2**

1. Test your figures from Practical 13.1 to see whether they appear significantly different from random numbers, on the basis of their frequency distribution.
2. Perform the same χ^2 test on your results to Practical 13.3. Are 'favourite' numbers random? (As you might not have 100 numbers in your final sample here, you will have to calculate your expected frequencies by dividing the total frequency for the sample by 10.)

Section **13.5** **Testing a Binomial Probability Model**

Practical **13.4**   **Mixing of Boys and Girls at Primary School**

Problem: **Do boys and girls at primary school mix together freely at lunch-times or do they prefer to sit with children of the same sex?**

Method

Obtain permission to go into a primary-school dining room and count the number of boys and girls at each table. In most primary schools the lunch-time assistants or teachers on duty ensure that children fill up the tables as they come in for lunch. This will help to make the seating arrangements more random than if the children are allowed to choose their own seats.

You must ensure that each table has the *same* number of seats at each table, and only include full tables in your sample. You will need to collect data for 50 tables (at least); so you may need to visit more than one school. Alternatively, you may go back to the same school on different days. Write down either the number of girls or the number of boys at each table.

Our Results

In the primary school used for our survey, children sit 8 to a table and have to fill up the tables as they arrive in the dining room. There are more boys than girls in the school.

The Null Hypothesis

Children are seated at lunch tables independently of their sex, and the distribution of boys (or girls) at each table can be modelled by a binomial distribution.

Under H_0, either of the following holds.

1. Children arrive randomly at the dining room and the order of boys and girls is determined by chance.
2. Children arrive with their friends—but their choice of friends is independent of sex.

The binomial distribution would be a valid model as each seat may be taken as a Bernoulli trial in which either a boy or a girl sits down. Each table (with 8 seats) can be taken as 8 Bernoulli trials so that $n = 8$. In a binomial situation, each trial should be independent. It is arguable as to whether that is so here.

Table 13.7 is the frequency table for the number of boys per table.

Table 13.7 Frequency table for the number of boys per table

Number x of boys per table	Frequency f
0	5
1	0
2	8
3	6
4	7
5	7
6	5
7	2
8	9
Total	49

We are testing whether the binomial distribution is a useful model in this situation. Each table can be viewed as 8 Bernoulli trials where either a boy or a girl sits in the seat. Thus, $n = 8$, but we do not know the value of p (probability that a boy sits down). The value of p is not $1/2$ as the numbers of boys and girls in the school are not equal. There is no theoretical value for p in this situation, and so we estimate it from our sample:

$$\text{estimate of } p \text{ (the proportion of boys)} = \frac{\text{total number of boys}}{\text{total number of children}}$$

The total number of boys (Table 13.8) may be found by multiplying each value for the number x of boys per table by its frequency f, i.e.

$$\text{total number of boys} = \sum fx$$

Therefore, number of boys $= 213$

As we have 49 tables each with 8 children, the total number of children is $49 \times 8 = 392$. Thus,

$$\text{estimate of } p = \frac{213}{392}$$

$$= 0.54 \text{ (to two decimal places)}$$

Table 13.8   Total number of boys

x	f	fx
0	5	0
1	0	0
2	8	16
3	6	18
4	7	28
5	7	35
6	5	30
7	2	14
8	9	72
Total	49	213

The Significance Test

H_0: $X \sim B(8, 0.54)$

H_1: X is not $B(8, 0.54)$

Calculation of Binomial Probabilities and Expected Frequencies

Calculation of binomial probabilities and expected frequencies are shown in Table 13.9.

The observed and expected frequencies are shown in Table 13.10.

Unfortunately, we have some very small expected frequencies. As χ^2 is calculated as $\sum[(O-E)^2/E]$, any term with a small expected value for E will tend to overinflate the value of χ^2, because observed frequencies must be integers. For this reason, classes (or cells) in the table which have low frequencies are combined with neighbouring cells as in Table 13.11.

As a guide, it is often suggested that expected frequencies below 5 should not be used, but this is a conservative number and could be taken as 4 or 3.

Table 13.9 Binomial probabilities and expected frequencies ($n = 8$; $p = 0.54$; $q = 1 - p = 0.46$)

x	$p(x)$		Expected frequency $49p(x)$
0	(0.46)	≈ 0.002	0.1
1	$8 \times (0.46)^7 \times (0.54)^1$	≈ 0.019	0.9
2	$\dfrac{8 \times 7}{2}(0.46)^6 \times (0.54)^2$	≈ 0.077	3.8
3	$\dfrac{8 \times 7 \times 6}{3 \times 2 \times 1}(0.46)^5 \times (0.54)^3$	≈ 0.182	8.9
4	$\dfrac{8 \times 7 \times 6 \times 5}{4 \times 3 \times 2 \times 1}(0.46)^4 \times (0.54)^4 \approx 0.267$		13.1
5	$56 \times (0.46)^3 \times (0.54)^5$	≈ 0.250	12.3
6	$28 \times (0.46)^2 \times (0.54)^6$	≈ 0.147	7.2
7	$8 \times (0.46)^1 \times (0.54)^7$	≈ 0.049	2.4
8	$(0.54)^8$	≈ 0.007	0.3

Table 13.10 Observed and expected frequencies

x	0	1	2	3	4	5	6	7	8
O	5	0	8	6	7	7	5	2	9
E	0.1	0.9	3.8	8.9	13.1	12.3	7.2	2.4	0.3

Table 13.11 Revised table

x	0, 1, 2	3	4	5	6, 7, 8
O	13	6	7	7	16
E	4.9	8.9	13.1	12.3	9.9

To Determine v, the Number of Degrees of Freedom

In our revised table, there were 5 cells for expected frequencies:

v = number of classes or cells − number of constraints

There is one constraint which determines that the total frequency must be 49. Also, in this example we have also had to estimate p from our sample. This has introduced another constraint. So $v = 5 - 2 = 3$. (For binomial models where there is a theoretical value for p, and p is not estimated from the sample data, this extra constraint does not exist. In that case, v = number of cells − 1 as there is just the 1 constraint for the total frequency.)

With $v = 3$, χ^2 is significant at the 5% level for values of $\chi^2 > 7.81$. Table 13.12 is the calculation of χ^2 arranged as a table.

Table 13.12   Calculation of $\chi^2 = \sum[(O-E)^2/E]$

O	E	$O-E$	$(O-E)^2$	$(O-E)^2/E$
13	4.9	8.1	65.61	13.39
6	8.9	-2.9	8.41	0.95
7	13.1	-6.1	37.21	2.84
7	12.3	-5.3	28.09	2.28
16	9.9	6.1	37.21	3.76
			Total	23.22

Our value for $\chi^2 = 23.22$ is significant at the 5% level (and also at the 1% level for which $\chi^2 > 11.34$).

We must conclude then at this level that our data do not fit the binomial model. The observed frequencies for the number of boys sitting at tables of 8 are significantly different from the expected frequencies generated by the model $B(8, 0.54)$. This is probably due to some lack of independence and there must be some sex preference in the choice of friends for children of primary-school age.

If you have access to a large nursery school, you may like to investigate whether the binomial model might be better suited to younger children.

Practical Follow-up 13.3

Analyse your data by following through all the steps in our working.

1. Estimate p from your sample.
2. Calculate expected frequencies from the binomial model (n is the number of seats at a dining table).
3. Combine any cells having expected frequencies less than 5, with adjacent cells.
4. Find $v = $ number of cells $- 2$ (for revised table).
5. Calculate χ^2.
6. Is your value of χ^2 significant?

Section 13.6   Testing a Poisson Model

If you have previously worked through Chapter 4, you will already have data available for testing in this section.

We are testing data collected on a motorway for the number of cars passing an observation point during 15 s intervals. Full details of how to organise the data collection are given in Chapter 4 together with other practical ideas, and a full explanation of the Poisson process.

Motorway Traffic: Our Results (Observed Frequencies)

We counted the number of vehicles passing our observation point (on the north-bound carriageway of the M1) for 76 intervals each of 15 s. Table 13.13 is the frequency table for the number of vehicles passing in a 15 s interval.

Table 13.13   Frequency table for the number of vehicles passing in 15 s intervals

x	f
0	1
1	4
2	8
3	6
4	17
5	18
6	13
7	3
8	3
9	1
10	1
11	1
12	0
Total	76

To Calculate the Expected Frequencies for a Poisson Model

The first step is to calculate the mean from the observed frequencies. This mean is then used as an estimate for μ, the mean of the Poisson distribution for which the expected frequencies are calculated. As we have used this estimate from our observed frequencies, this places an extra restriction on the number of degrees of freedom.

The mean of the above data is 4.6, and from this Poisson probabilities (Table 13.14) and the expected frequencies (Table 13.15) have been calculated. The working for this is shown in full detail in Chapter 4.

Again, several of the expected frequencies are below 5, and so the cells for $x = 0$, 1 and 2 are combined as are the cells for $x = 8$, 9, 10 and 11. Table 13.16 is the revised table.

The Significance Test

The final table has 7 cells for expected frequencies but there are two restrictions.

1. The total expected frequency must equal 76.
2. The mean for the Poisson model has been estimated from the observed frequencies.

So

v = number of cells or classes − number of restrictions

$$= 7 - 2$$

$$= 5$$

Testing at the 5% level, the critical value for χ_5^2 is 11.07.

The Null Hypothesis

$H_0: X \sim P(4.6)$

$H_1: X$ is not $P(4.6)$

Table 13.14 Expected frequencies calculated from a Poisson distribution with $\mu = 4.6$

x	Poisson probabilities $p(x) = e^{-4.6}\mu^x/x!$		Expected frequency $76p(x)$ (given to one decimal place)
0	$e^{-4.6}$	≈ 0.01005	0.8
1	$e^{-4.6} \times 4.6$	≈ 0.0462	3.5
2	$e^{-4.6} \times \dfrac{(4.6)^2}{2!}$	≈ 0.1063	8.1
3	$e^{-4.6} \times \dfrac{(4.6)^3}{3!}$	≈ 0.1631	12.4
4	$e^{-4.6} \times \dfrac{(4.6)^4}{4!}$	≈ 0.1875	14.3
5	$e^{-4.6} \times \dfrac{(4.6)^5}{5!}$	≈ 0.1725	13.1
6	$e^{-4.6} \times \dfrac{(4.6)^6}{6!}$	≈ 0.1323	10.1
7	$e^{-4.6} \times \dfrac{(4.6)^7}{7!}$	≈ 0.0869	6.6
8	$e^{-4.6} \times \dfrac{(4.6)^8}{8!}$	≈ 0.0500	3.8
9	$e^{-4.6} \times \dfrac{(4.6)^9}{9!}$	≈ 0.0255	1.9
10	$e^{-4.6} \times \dfrac{(4.6)^{10}}{10!}$	≈ 0.0118	0.9
11 and over	$1 - p(x \leqslant 10)$		0.5

Table 13.15 Observed and expected frequencies

x	0	1	2	3	4	5	6	7	8	9	10	11 or more
Observed	1	4	8	6	17	18	13	3	3	1	1	1
Expected	0.8	3.5	8.1	12.4	14.3	13.1	10.1	6.6	3.8	1.9	0.9	0.5

Table 13.16 Revised table for observed and expected frequencies

x	0, 1, 2	3	4	5	6	7	8 or more
O	13	6	17	18	13	3	6
E	12.4	12.4	14.3	13.1	10.1	6.6	7.1

Calculation of χ^2

$$\chi^2 = \sum \frac{(O-E)^2}{E}$$

Using the above formula, we have calculated our total for χ^2, setting out the working in the form of a table (Table 13.17).

Table 13.17 Calculation of χ^2

x	O	E	$O-E$	$(O-E)^2$	$(O-E)^2/E$
0, 1, 2	13	12.4	0.6	0.36	$\frac{0.36}{12.4} \approx 0.03$
3	6	12.4	-6.4	40.96	$\frac{40.96}{12.4} \approx 3.30$
4	7	14.3	2.7	7.29	$\frac{7.29}{14.3} \approx 0.51$
5	18	13.1	4.9	24.01	$\frac{24.01}{13.1} \approx 1.83$
6	13	10.1	2.9	8.41	$\frac{8.41}{10.1} \approx 0.83$
7	3	6.6	-3.1	12.96	$\frac{12.96}{6.6} \approx 1.96$
8, 9, 10, 11	6	7.1	-1.1	1.21	$\frac{1.21}{7.1} \approx 0.17$
				Total	8.63

This total is less than the critical value of 11.07; so our observed and expected frequencies are not significantly different at the 5% level. We can conclude that our data collected on the M1 motorway do conform reasonably well with the frequencies predicted by a Poisson model.

Low Expected Frequencies

In our working in this section, we adhered very strictly to the 'rule' that cells with expected frequencies below 5 should be combined. However, it is possible to be reasonably flexible about this, as we shall show here.

By combining only $x = 0$ and $x = 1$, we have a total expected frequency of 4.3, and it could be argued that it is not necessary to combine these again with $x = 2$ (for which the expected frequency is 8.1). At the other end of the table, we could combine the values $x = 9$, 10 and 11, giving a total expected frequency of 3.3, and $x = 8$ could stand on its own.

Table 13.18 is the table of observed and expected frequencies that we obtain.

Table 13.18 Observed and expected frequencies

x	0, 1	2	3	4	5	6	7	8	9, 10, 11
O	5	8	6	17	18	13	3	3	3
E	4.3	8.1	12.4	14.3	13.1	10.1	6.6	3.8	3.3

This arrangement gives us two more cells than the previous solution (and, as a result, 2 more degrees of freedom).

Now,

$v = 9 - 2$

$v = 7$

Testing at the 5% level, the critical value for χ^2 is 14.07. Recalculating χ^2 on the basis of our new table for observed and expected frequencies, our total comes to 8.73 (only 0.1 different from our previous total). However, we have gained two degrees of freedom and the critical value required for χ^2 has increased to 14.07 (compared with 11.07 for χ_5^2).

Provided that expected frequencies are not allowed to go below 3 or 4, our end results will not change very much from those obtained by rigidly applying the '5 rule'.

Practical Follow-up 13.4

Analyse your results from your practical (which you may have completed some time ago when working through Chapter 4) by following through all the steps in our working.

1. Find the mean of your observed frequency distribution.
2. Calculate the expected frequencies from the Poisson model.
3. Combine any cells having expected frequencies less than 5 with adjacent cells.
4. Find the number of degrees of freedom

   $v =$ number of cells $- 2$

5. Find the significant value for χ^2 from tables.
6. Calculate χ^2.
7. Is your value of χ^2 significant? What are your conclusions?

Section 13.7   Testing the Normal Distribution as a Model

The data that we test in this section were collected as a practical assignment in Chapter 5. Several alternative practicals are suggested in that chapter and full instructions are given for the collection of suitable data.

We measured both the widths and the lengths of 100 laurel leaves. Table 13.19 shows the frequency distribution for the lengths of these leaves.

Table 13.19   Lengths of 100 laurel leaves

Length (mm)	Frequency
80–89	1
90–99	0
100–109	3
110–119	10
120–129	19
130–139	21
140–149	23
150–159	11
160–169	8
170–179	1
180–189	2
190–199	1

The Normal Distribution Model

In order to calculate the expected frequencies generated by a normal distribution model, work through the following steps.

1. *If necessary, estimate μ and σ^2.* In order to test whether a sample may have come from a normal distribution, we need to know μ and σ^2. These may be given, if we are testing whether the sample comes from a known normal distribution. (In this case, there is only one constraint on the number of degrees of freedom and that is the total frequency.)

   More generally, we do not know μ or σ^2; so we have to estimate them from the sample using

$$\bar{x} = \frac{\sum_i f_i x_i}{\sum_i f_i}$$

   for μ and

$$s^2 = \frac{\sum_i f_i (x_i - \bar{x})^2}{\sum_i f_i - 1}$$

   for σ^2. In this case, there will be three constraints when we calculate the number of degrees of freedom.

   For the data given here, $\bar{x} = 138.1$ and $s^2 = 342.46$.

2. *Find the true class limits for grouped data in the table.* As our measurements were taken to the nearest millimetre, we could be incorrect by 0.5 mm either way. Our true class intervals are

   79.5 – 89.5 mm (just under)

   89.5 – 99.5 mm (just under)

   99.5 – 109.5 mm (just under)

3. The upper limit for each class interval is expressed as a standardised normal variate (i.e. its difference from the mean is divided by the standard deviation):

$$z = \frac{x - \mu}{\sigma}$$

   We use

$$z = \frac{x - \bar{x}}{s}$$

   To express 89.5 as a standard normal variable, we calculate

$$z = \frac{89.5 - 138.1}{18.5}$$

$$= -2.63$$

   (as $\bar{x} = 138.1$ and $s = 18.5$).

   Find the area to the left of each class interval upper limit ($\Phi(z)$ in standard normal tables which gives the value of the distribution function or cumulative probability). For $x = 89.5$, $z = -2.63$ and $\Phi(z) = 0.004\,24$.

5. Calculate the expected cumulative frequency by multiplying $\Phi(z)$ by the total frequency (which in this case is 100).

6. The expected frequencies can be easily found from the cumulative frequencies.
   Tables 13.20, 13.21 and 13.22 give the full calculations for finding the expected

Table 13.20   Cumulative frequencies

True class limits	Upper limit	z	$\Phi(z)$	Cumulative frequency $100\Phi(z)$	Expected frequency
−79.5	79.5	−3.17	0.000 76	0	0
79.5−89.5	89.5	−2.63	0.004 27	0.4	0.4
89.5−99.5	99.5	−2.09	0.018 31	1.8	1.4
99.5−109.5	109.5	−1.55	0.060 6	6.1	4.7
109.5−119.5	119.5	−1.01	0.156 2	15.6	9.5
119.5−129.5	129.5	−0.46	0.322 8	32.3	16.7
129.5−139.5	139.5	0.08	0.531 9	53.2	20.9
139.5−149.5	149.5	0.62	0.732 4	73.2	20.0
149.5−159.5	159.5	1.16	0.877 0	87.7	14.5
159.5−169.5	169.5	1.70	0.9554	95.5	7.8
169.5−179.5	179.5	2.24	0.987 45	98.7	3.2
179.5−189.5	189.5	2.78	0.997 28	99.7	1.0
189.5−199.5	199.5	3.32	0.999 55	99.9	0.2
Over 199.5	∞		1.0	100.0	0.1

Table 13.21   Observed and expected frequencies

Upper limit for x	Observed frequency O	Expected frequency E	
79.5	0	0	Combine these
89.5	1	0.4	groups (total
99.5	0	1.4	expected
109.5	3	4.7	frequency, 6.5)
119.5	10	9.5	
129.5	19	16.7	
139.5	21	20.9	
149.5	23	20.0	
159.5	11	14.5	
169.5	8	7.8	
179.5	1	3.2	Combine these
189.5	2	1.0	groups (total
199.5	1	0.2	expected
	0	0.1	frequency, 4.5)

Table 13.22   Revised table for observed and expected frequencies

Upper limit for x	Observed frequency	Expected frequency
109.5	4	6.5
119.5	10	9.5
129.5	19	16.7
139.5	21	20.9
149.5	23	20.0
159.5	11	14.5
169.5	8	7.8
∞	4	4.5

frequencies from the normal distribution $N(138.1, 342.46)$. Note that we have first found the area to the left of our first group starting at 79.5. So

$$z = \frac{79.5 - 138.1}{18.5}$$

$$= -3.17$$

In this revised table, cells with very low frequencies have been combined.

The Significance Test

To calculate v, the number of degrees of freedom, we must determine the number of restrictions. Since the total of the expected frequencies must equal 100, and we have estimated two parameters μ and σ^2 from the observed frequencies, there are 3 restrictions. Thus,

$v =$ number of cells $-$ number of restrictions

$= 8 - 3$

$= 5$

Testing at the 5% level, the critical value for χ^2 is 11.07.

The Null Hypothesis

H_0: $X \sim N(138.1, 342.46)$

H_1: X is not $N(138.1, 342.46)$

Calculation of χ^2

Table 13.23 gives our calculations for χ^2.

Table 13.23 Calculation of χ^2

Upper limit for x	O	E	$O - E$	$(O - E)^2$	$(O - E)^2/E$
109.5	4	6.5	−2.5	6.25	0.962
119.5	10	9.5	0.5	0.25	0.026
129.5	19	16.7	2.3	5.29	0.317
139.5	21	20.9	0.1	0.01	0.000
149.5	23	20.0	3.0	9.0	0.45
159.5	11	14.5	−3.5	12.25	0.845
169.5	8	7.8	0.2	0.04	0.005
∞	4	4.5	0.5	0.25	0.056
				Total	2.661

Therefore,

total for $\chi^2 = \sum \dfrac{(O - E)^2}{E}$

This value of χ^2 is not significant at the 5% level; therefore we can conclude that our normal data conform very closely to the model that we have chosen, i.e. a normal distribution $N(131.8, 342.46)$.

Practical Follow-up 13.5

Find the data that you collected as a practical for Chapter 5. Work through the stages set out in this section to establish whether your data fit closely to a normal distribution model.

Section 13.8   Testing Distributions in Contingency Tables

Practical 13.5   A Comparison of the Distribution of Two Industries Among the Postal Districts in a Large Town

Background

In any large town or city, there are likely to be zones of different types of industry. In the centre of a city are generally found the banks, shops and offices together with business services such as printing, photocopying and computer services. This area may well also contain small workshops for traditional craft industries dating back to the beginnings of the Industrial Revolution. Further out from the city centre close to railway lines or canals may be found heavy engineering or textile industries established in the later part of the nineteenth century or early twentieth century. Newer industries are generally found on industrial estates built on the outskirts of towns.

Our Problem

We wanted to establish whether the distribution of cutlery manufacturers is significantly different from the distribution of heavy engineering in Sheffield (the cutlery industry is a very old craft industry for which Sheffield has always been renowned).

You may adopt your hypothesis to suit the city or town you are studying. The town you choose should be large, have a variety of industries and have a *Yellow Pages* directory.

Materials Required

1. *Yellow Pages* for the town or city or your choice.
2. A postcodes book from the Post Office—this also contains a useful postcodes map.
3. A street map may be helpful.

Collecting your Data

Before you start, consult the postcodes book to decide which postcodes lie within the city boundary. Discard all postcodes for surrounding towns and country areas.

Find out about the traditional industries of your city or town. Make a list of these and decide which to include in your study. Decide on a list of newer industries—you might choose food processing, confectionery and soft-drinks manufacturers as a group together, or computers and microelectronics or light engineering.

Construct a table with a vertical column for each postcode within the town and with a horizontal row for each minor industry within the group that you have chosen.

Our Study: City of Sheffield Postal Districts

Figure 13.4 is a map showing Sheffield postcodes. Most codes lie within the city boundary except for S61 which is a neighbouring town, Rotherham.

Most of S30 consists of small villages which have gradually become suburbs of the city as it has spread outwards. As we collected our data, we found very few industries in S30, and for that reason S30 does not appear in later analysis, nor do other suburban residential districts S5, S6, S7, S10, S11, S12, S14 and S17. Note that these residential areas are predominantly to the west side of Sheffield which is very different in character to the industrial east side.

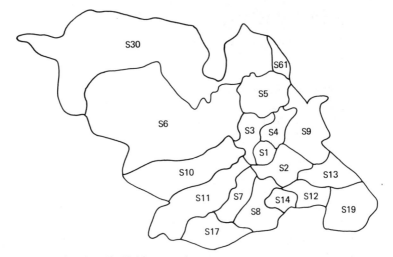

Figure 13.4 Map showing Sheffield postcodes

Our Results

By consulting the *Yellow Pages*, each cutlery manufacturer was identified and recorded under the correct postcode (all postcodes are given at this stage) (Table 13.24).

Table 13.24 Distribution of cutlery manufacturers

Postal district	S1	S2	S3	S4	S5	S6	S7	S8	S9	S10
Number of manufacturers	47	10	27	8	0	2	0	2	2	0

Post district (continued)	S11	S12	S13	S14	S30
Number of manufacturers (*continued*)	0	0	2	0	0

Next we found data for the following group of industries which were to be amalgamated under the heading of 'heavy engineering': alloy steel; bearings and bolts; castings; engineering; forgings and stampings; foundries. Table 13.25 gives a full set of data for all postal districts.

Table 13.25 Heavy engineering

Postal district	S1	S2	S3	S4	S5	S6	S7	S8	S9	S10	S11	S12	S13	S14	S30
Alloy steel	0	0	1	1	0	1	1	0	5	1	0	0	0	0	0
Bearings and bolts	0	1	2	1	0	4	0	0	5	0	0	0	0	0	0
Castings	2	0	4	1	0	0	0	1	6	0	0	0	1	0	0
Engineering	9	9	21	11	0	11	0	4	37	2	4	2	3	0	3
Forgings and stampings	1	3	2	0	0	7	0	1	8	0	0	0	0	0	0
Foundries	3	1	3	5	0	1	0	0	1	0	0	0	0	0	0
Total heavy engineering	15	11	34	12	0	24	1	6	62	3	4	2	4	0	3

As some residential areas did not contain any factories in these categories (or very few), they were not included in our final table, Table 13.26.

These figures are the 'observed frequencies' for the distribution of the industries we have chosen.

Table 13.26   Distributions of cutlery manufacturers and heavy engineering in Sheffield

Postal district	S1	S2	S3	S4	S6	S8	S9	S13
Number of cutlery manufacturers	47	10	27	3	2	2	2	2
Number of engineering units	15	11	34	12	24	6	62	4

The Null Hypothesis

The null hypothesis is that there is no difference between the distributions of manufacturing units for the two industries. Under that hypothesis the distributions should be such that the manufacturers in each postal district are split between cutlery and engineering in the same proportions as the overall totals. This idea should become clearer as we calculate the expected frequencies under H_0.

Firstly, calculate the column totals (the total number of factories in each postal district) and the row totals (the total number of factories for each type of industry (Table 13.27).

Table 13.27   Observed frequencies

Postal district	S1	S2	S3	S4	S6	S8	S9	S13	Total
Cutlery manufacturers	47	10	27	3	2	2	2	2	95
Engineering units	15	11	34	12	24	6	62	4	168
Total	62	21	61	15	26	8	64	6	263

Altogether there are 95 cutlery manufacturers and 168 heavy engineering units. So the expected distribution of factories under H_0 is found by splitting the total for each postal district in the ratio 95 to 168.

For example, in Sheffield 1, there are 62 factories altogether. Therefore,

$$\text{expected number of cutlery factories} = 62 \times \frac{95}{263}$$

$$= 22.4 \text{ (to one decimal place)}$$

and

$$\text{expected number of heavy engineering units} = 62 \times \frac{168}{263}$$

$$= 39.6 \text{ (to one decimal place)}$$

The other frequencies are found in the same way. The following formula may be useful:

$$\text{expected frequency} = \text{column total} \times \frac{\text{row total}}{\text{grand total}}$$

Table 13.28 is our completed table of expected frequencies.

Table 13.28 Expected frequencies

Postal district	S1	S2	S3	S4	S6	S8	S9	S13	Total
Cutlery manufacturers	22.4	7.6	22.0	5.4	9.4	2.9	23.1	2.2	95
Heavy engineering units	39.6	13.4	39.0	9.6	16.6	5.1	40.9	3.8	168
Total	62	21	61	15	26	8	64	6	263

At this stage, you may find that some of your expected frequencies are too small. This means that you may have to combine postal districts (provided they are adjacent geographically). We decided to combine S13 and S2. Tables 13.29 and 13.30 are our revised tables for observed and expected frequencies.

Table 13.29 Observed frequencies

Postal district	S1	S2 and S13	S3	S4	S6	S8	S9	Total
Cutlery manufacturers	47	12	27	3	2	2	2	95
Heavy engineering units	15	15	34	12	24	6	62	168
Total	62	27	61	15	26	8	64	263

Table 13.30 Expected frequencies

Postal district	S1	S2 and S13	S3	S4	S6	S8	S9	Total
Cutlery manufacturers	22.4	9.8	22.0	5.4	9.4	2.9	23.1	95
Heavy engineering units	39.6	17.2	39.0	9.6	16.6	5.1	40.9	169
Total	62	27	61	15	26	8	64	263

The Number of Degrees of Freedom

In our final table for the expected frequencies, we have 14 cells to be filled, but we do not have 14 degrees of freedom. Note that each column total and each row total is the same as for the observed frequencies. These are constraints on our model and reduce our number of degrees of freedom.

For a table such as this (which is known as a *contingency table*) with m rows and n columns, the number of degrees of freedom is calculated from

$$v = (m - 1)(n - 1)$$

In our contingency table we have 2 rows and 7 columns so that

$$v = (2 - 1)(7 - 1)$$
$$= 6$$

Of our 14 cells for expected frequencies, only 6 of our cells can be filled independently; all the others will be determined by the row and column totals.

By consulting statistical tables, we find that, testing our hypothesis at the 5% level, χ^2 must be greater than 12.59 for a significant result. Should we wish to test at the 1% level, the significant value for χ^2 is 16.81 (i.e. if $\chi^2 > 16.81$ there is a probability of less than 0.01 that the observed frequencies are random samples from two equal distributions).

Calculating χ^2

As before, the χ^2 total is calculated as

$$\chi^2 = \sum \frac{(O - E)^2}{E}$$

Our working is given in Table 13.31.

Table 13.31   Calculation of χ^2

	Postal district	O	E	$O - E$	$(O - E)^2$	$(O - E)^2/E$
Cutlery	S1	47	22.4	24.6	605.16	27.02
	S2 and S13	12	9.8	2.2	4.84	0.49
	S3	27	22.0	5.0	25.0	1.14
	S4	3	5.4	−2.4	5.76	1.07
	S6	2	9.4	−7.4	54.76	5.83
	S8	2	2.9	−0.9	0.81	0.28
	S9	2	23.1	−21.1	445.21	19.27
Engineering	S1	15	39.6	24.6	605.16	15.28
	S2 and S13	15	17.2	−2.2	4.84	0.28
	S3	34	39.0	−5.0	25.0	0.64
	S4	12	9.6	2.4	5.76	0.60
	S6	24	16.6	7.4	54.76	3.30
	S8	6	5.1	0.9	0.81	0.16
	S9	62	40.9	21.1	445.21	10.89
					Total	86.25

Our total for χ^2 is highly significant, and we can conclude that the locations of cutlery manufacture and heavy engineering are very different in Sheffield.

Practical Follow-up 13.6

1. During the course of our investigation into the location of industry in Sheffield, we also collected the data in Table 13.32 from tool makers and cutlery manufacturers. Are the distributions for these two industries significantly different?
2. Conduct a similar study to ours on two types of industry in the town or city of your choice. Conduct a χ^2 test on your data to decide whether the industries are differently located.

Table 13.32   Distributions of cutlery manufacturers and tool makers in Sheffield

Postal district	S1	S2	S3 and S4	S6	S9
Cutlery manufacturers	47	10	30	2	2
Tool makers	10	8	19	10	20

The Relationship Between the χ^2 and Normal Distributions

In order to show the relationship between χ^2 and normal distributions, we start with a simple contingency table in which frequencies may fall into either one of two groups only. (You may like to think of group A as 'success' and group B as 'failure'.)

Observed Frequencies

	Frequency
Group A	n_1
Group B	n_2
Total	n

The observed frequencies are n_1 in group A and n_2 in group B, with a total frequency $n_1 + n_2 = n$. To calculate the expected frequencies, we need to define the probability of being in group A as p, and of being in group B as q (so that $p + q = 1$). The expected frequencies will therefore be np for group A and nq (or $n(1 - p)$) for group B.

Expected Frequencies

	Frequency
Group A	np
Group B	nq
Total	n

We then calculate χ^2 for the observed and expected frequencies. With only 2 cells in the table and 1 constraint (the total frequency n), the number of degrees of freedom is 1.

$$\chi_1^2 = \frac{(n_1 - np)^2}{np} + \frac{(n_2 - nq)^2}{nq}$$

$$= \frac{(n_1 - np)^2}{np} + \frac{[n_2 - n(1 - p)]^2}{nq}$$

Now $n_1 + n_2 = n$. So $n_2 - n = -n_1$. Therefore,

$$\chi_1^2 = \frac{(n_1 - np)^2}{np} + \frac{(n_2 - n - np)^2}{nq}$$

$$= \frac{(n_1 - np)^2}{np} + \frac{-n_1 - np^2}{nq}$$

$$= \frac{(n_1 - np)^2}{np} + \frac{(np - n_1)^2}{nq}$$

Note that $(n_1 - np)^2 = (np - n_1)^2$ so that

$$\chi_1{}^2 = \frac{(n_1 - np)^2}{np} + \frac{(n_1 - np)^2}{nq}$$

Using a common denominator,

$$\chi_1{}^2 = \frac{q(n_1 - np)^2 + p(n_1 - np)^2}{npq}$$

$$= \frac{(p + q)(n_1 - np)^2}{npq}$$

As $p + q = 1$,

$$\chi_1{}^2 = \frac{(n_1 - np)^2}{npq}$$

Also, we can think of n_1 as being binomially distributed, as there are only two outcomes: group A (success) and group B (failure) for each of n trials. Thus $n_1 \sim B(n, p)$.

Provided that n is large, the distribution of n can be approximated by a normal distribution with mean np and standard deviation $\sqrt{npq}$. Thus, $(np - n_1)/\sqrt{npq}$ is a standardised normal variate Z (where $Z \sim N(0, 1)$) and we also have

$$\chi_1{}^2 = \frac{(np - n_1)^2}{npq}$$

So

$$\chi^2 = Z^2$$

If X is distributed as $N(0, 1)$, then X^2 is distributed as χ^2 with 1 degree of freedom. If $X_1, X_2, X_3, \ldots, X_n$ are independent $N(0, 1)$ distributions, then

$$\sum_{i=1}^{n} X_i{}^2 \text{ is distributed as a } \chi^2 \text{ random variable with } n \text{ degrees of freedom.}$$

Yates' Correction

Yates' correction is used only for small contingency tables where $v = 1$ (these will generally be 2×2 contingency tables or possibly 2×1 as used in our proof just given). In these contingency tables we are comparing two proportions. The sampling distribution of possible values of $\sum[(O - E)^2/E]$ generated by a 2×2 table will be a discrete distribution with a finite number of values. This is because the observed frequencies from which we make our calculations must be whole numbers.

In contrast, the χ^2 distribution is continuous. To improve the agreement between the χ^2 distribution and the sampling distribution of $\sum[(O - E)^2/E]$, we use a continuity correction known as Yates' correction. This involves reducing each absolute difference of $|O - E|$ by $1/2$ before squaring it. Thus,

$$\chi^2 = \sum \frac{(|O - E| - 0.5)^2}{E}$$

Chapter 14 The Exponential Distribution

Section 14.1 A Survey

Our nearest petrol station has 4 petrol pumps and, although the half-hour morning and evening rush-hour periods are particularly busy, there is a constant stream of customers all day. The manageress is wondering whether it would be worthwhile installing 1 or 2 more petrol pumps. It is a self-service station and she estimates that customers take on average 3 or 4 min to serve themselves and to pay for their petrol, provided that they do not have to queue.

Her son, who studies statistics, offers to conduct a survey and spends about $1\frac{1}{2}$ hours one morning (from 10.00 am onwards) out on the forecourt. He decides to record the number of cars arriving in each 1 min interval and obtains the following results.

Number of cars arriving in a 1 min interval	Frequency
0	29
1	38
2	15
3	6
4 or more	0

From these he calculates the mean to be 0.977. He says that this shows that, as cars arrive roughly one per minute on average, there should be no problem. 'But...,' says his mother, 'if we look at your figures, there are some minute intervals where no cars arrive at all, and others where several cars arrive together. I think what I would really like to know is: how long is it between the arrival of one car and the next?'

Section 14.2 Practicals

Practical 14.1   Times between Arrivals

For a period of about an hour, record the intervals in seconds between any of the following.

1. Cars arriving at a petrol station or motorway service station.
2. Customers arriving at a bank, Post Office or shop.
3. Telephone calls arriving at a switchboard.

Alternatively, if you record the actual times of arrivals, you will be able to analyse your results in two ways—either in terms of the number of arrivals in each 60 s interval or in terms of the lengths of time between arrivals.

Practical 14.2   Geiger Counter

Record the time between clicks on a Geiger counter measuring background radiation (a radioactive source is not required).

Practical 14.3   Letter 'a' in a Piece of Prose

Count and record the number of letters between successive occurrences of the letter 'a' in a piece of prose.

Section 14.3   Our Results

The results in Table 14.1 and Figure 14.1 show the distribution of the lengths of time between the arrivals of consecutive cars at the same petrol station for the same period of time as the results shown in Section 14.1.

Table 14.1   Results for Practical 14.1

Time intervals between car arrivals (s)	Frequency
0–	32
30–	24
60–	10
90–	9
120–	6
150–	0
180–	1
210–	2
240–	2
270–	0
300–	1

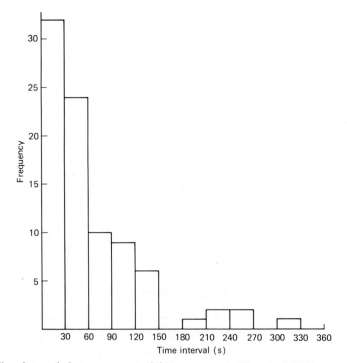

Figure 14.1   Time intervals between cars arriving at a garage (Practical 14.1)

The raw results are given at the end of the chapter, in Table 14.5.

The mean time interval is found to be just under 61 s which certainly ties in with the previous result of just under 1 car arriving on average in each 1 min period.

Is there a connection between the two surveys? If we can find a model for the number of cars arriving per minute, can we use this to help us to derive a model for the time intervals between cars?

The First Survey Again: The Number of Cars Arriving in 1 min Intervals

Table 14.2 and Figure 14.2 shows the results of the first survey again.

Table 14.2   Results of the first survey

Number of cars arriving in a 1 min interval	Frequency
0	29
1	38
2	15
3	6
4 or more	0

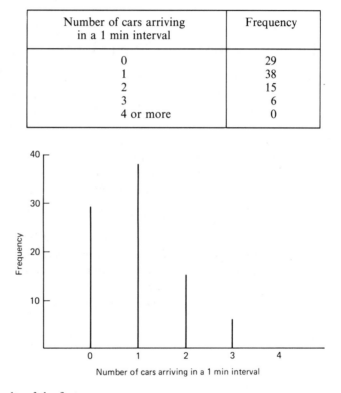

Number of cars arriving in a 1 min interval

Figure 14.2   Results of the first survey

The line graph in Figure 14.2 suggests that possibly the Poisson distribution might be a suitable model for the number of cars arriving in each 1 min interval.

The mean and variance for these data are 0.977 and 0.772, respectively, and the fact that these are similar in value gives further support to our choice of model. If μ is taken as 1 for easier calculation, the corresponding theoretical Poisson probabilities and expected frequencies can be calculated (Table 14.3).

These are very close indeed to the observed frequencies obtained in the survey (and, if a χ^2 test is made, the result is not significant, as shown at the end of the chapter). Thus the Poisson distribution does seem to provide us with a suitable model for the number of cars arriving in each minute.

Table 14.3 Theoretical Poisson frequencies for $\mu = 1$

x	$p(x)$	Expected frequency $88p(x)$
0	$e^{-1} \approx 0.368$	32.4
1	$e^{-1} \approx 0.368$	32.4
2	$\dfrac{e^{-1}}{2!} \approx 0.184$	16.2
3	$\dfrac{e^{-1}}{3!} \approx 0.061$	5.4
4 or more	≈ 0.019	1.6

Section 14.4 Finding a Probability Model for the Time Intervals Between Cars

We have found that the number of cars arriving in a particular time interval is a discrete variable which can be modelled by a Poisson probability distribution. What we want to find now is a probability model for the time in seconds between one car and the next. This will enable us to answer questions such as what is the probability that two cars will arrive less than a minute apart and so on. Time is a continuous variable and so we need to find a continuous function for our probability model.

If we go back to our Poisson model for the number of cars arriving per minute, we can use that to help us. The average number of cars arriving in a minute is approximately 1 (thus the average number of cars arriving in a second is $1/60$). If the rate of arrival of cars is, on average, λ per second, the average number arriving in t will be λt. The number of cars arriving in t s will then follow a Poisson distribution with mean λt.

So the probability of no cars arriving in t seconds is $e^{-\lambda t}$.

Now all the above is true for any value of t. Let T be the time between two successive cars. Then $P(T > t)$ is the same as the probability that there are no cars in the time interval t. They are only different ways of describing the same event.

So

$$P(T > t) = e^{-\lambda t}$$

and

$$P(T < t) = 1 - e^{-\lambda t}$$

Let the probability density function for T be $f(t)$ and the distribution function be $F(t)$. We have

$$F(t) = \int_0^t f(x)\,dx = 1 - e^{-\lambda t} \qquad 0 \leqslant t < \infty$$

Differentiate to get

$$f(t) = \lambda e^{-\lambda t}$$

or, with the more usual variable x,

$$f(x) = \lambda e^{-\lambda x}$$

This probability density function is known as the exponential distribution and its range is $0 < x < \infty$ and λ is the average number of cars arriving per second, i.e. the mean of the Poisson distribution which we started with.

Calculating Theoretical Exponential Frequencies

From our raw data, the average length of time between cars is 61 s (rather than 60 s).

We can calculate the theoretical probabilities and expected frequencies for an exponential distribution with $\lambda = 1/61$ and compare these with our actual results.

It is easier, in this instance, to use the (cumulative) distribution function

$$F(x) = 1 - e^{-x/61}$$

Cumulative probabilities can be calculated by substituting the value for x at the end of each time interval. Probabilities for each separate time interval are then obtained by subtraction, and expected frequencies can be calculated by multiplying by 87 (Table 14.4).

Table 14.4   Probabilities, cumulative probabilities and expected frequencies

Class upper boundary (s)	Cumulative probability $F(x) = 1 - e^{-x/61}$	Probability $p(x)$	Expected frequency $87p(x)$
−30	0.388	0.388	33.8
−60	0.626	0.288	25.1
−90	0.771	0.145	12.6
−120	0.860	0.089	7.7
−150	0.914	0.054	4.7
−180	0.948	0.034	3.0
−210	0.968	0.020	1.7
−240	0.980	0.012	1.1
−270	0.988	0.008	0.7
−300	0.992	0.004	0.3
−330	0.995	0.003	0.3
−360	0.997	0.002	0.2

Comparison of Our Survey Results with the Exponential Model

A comparison is given in Figure 14.3.

From this histogram, it can be seen that the theoretical frequencies are very close to the actual frequencies obtained in the survey. (If a χ^2 test is done, the value of χ^2 obtained is not significant at the 5% level, as shown at the end of the chapter.) Thus the exponential distribution does appear to provide a close model for the length of time between consecutive arrivals.

Practical Follow-up 14.1

For your survey results, carry out the following procedure.

1. If you have recorded the time intervals between arrivals, draw up a frequency table based on suitable class intervals (giving approximately 10 class intervals across the range).

   Alternatively, if you have recorded arrival times, subtract each time from the next to calculate the intervals between arrivals. Then draw up a frequency table to summarise these results.

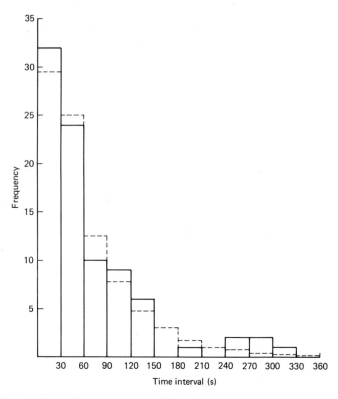

Figure 14.3 Comparison of our survey results with the exponential model: ———, observed frequency; –––, theoretical expected frequency

2. Calculate the mean and variance for the time between arrivals.
3. Taking $\lambda = 1/($this mean value$)$, calculate the theoretical probabilities for the exponential distribution and the expected frequencies.
4. Compare your results with the theoretical expected frequencies (conduct a χ^2 test if you have studied that chapter remembering that $v =$ number of classes $- 2$ as we have 2 restrictions: a set total for the expected frequencies and λ estimated from the data).

Section **14.5** The Theoretical Mean and Variance of the Exponential Distribution

A continuous random variable X has an exponential distribution if it has the probability density function

$$f(x) = \begin{cases} \lambda e^{-\lambda x} & x \geq 0 \qquad \lambda > 0 \\ 0 & x < 0 \qquad \lambda > 0 \end{cases}$$

For $\lambda = 1/61$ as in our example, the curve for the graph of the function is shown in Figure 14.4.

The mean μ is found by integrating $\displaystyle\int_0^\infty xf(x)\,dx$:

$$\mu = E(x) = \int_0^\infty x\lambda e^{-x}\,dx$$

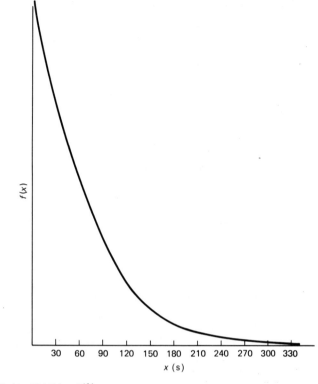

Figure 14.4 $f(x) = (1/61)\,e^{-x/61}$

Integrating by parts, we get

$$\int_0^\infty x\lambda\,e^{-\lambda x} = \left[-x\,e^{-\lambda x} \right]_0^\infty - \int_0^\infty -e^{-\lambda x} \times 1\,dx$$

$$= \left[-x\,e^{-\lambda x} \right]_0^\infty - \left[\frac{1}{\lambda}e^{-\lambda x} \right]_0^\infty$$

$$= 0 - \left(0 - \frac{1}{\lambda} \right)$$

$$= \frac{1}{\lambda}$$

Thus the average length of time between accidents is $1/\lambda$, where λ is the mean of the Poisson distribution used to model the number of cars per second and $1/\lambda$ is the mean of the exponential distribution for the length of time between successive cars. For our two surveys, we obtained

mean number of cars per minute = 0.977

mean length of time between cars = 61 s

$$= 1.02\ \text{min}$$

Thus, $\lambda = 0.977$ and $1/\lambda = 1.02$. (This is intuitively justified on the basis that, if arrivals were, say, on average 5 per second, then average time between arrivals would be $1/5$ s.)

Note that the two surveys were conducted together over the same period of time. To find the variance, we need $\int_0^\infty x^2 f(x)\, dx - (\text{mean})^2$:

$$\text{variance} = \int_0^\infty x^2 \lambda\, e^{-\lambda x}\, dx - \left(\frac{1}{\lambda}\right)^2$$

Integrating by parts as before, we get

$$\text{variance} = \left[-x^2\, e^{-x} \right]_0^\infty - \int_0^\infty -2x\, e^{-\lambda x}\, dx - \left(\frac{1}{\lambda}\right)^2$$

Integrating again by parts,

$$\text{variance} = \left[-x^2\, e^{-\lambda x} \right]_0^\infty - \left[2x\frac{1}{\lambda}\, e^{-\lambda x} \right]_0^\infty - \left[2\frac{1}{\lambda^2}\, e^{-\lambda x} \right]_0^\infty - \frac{1}{\lambda^2}$$

$$= 0 - 0 + \frac{2}{\lambda^2} - \frac{1}{\lambda^2}$$

$$= \frac{1}{\lambda^2}$$

Thus the variance is $1/\lambda^2$.

The Median

The median is the value $x = M$ such that

$$F(M) = \frac{1}{2}$$

$$1 - e^{-\lambda M} = \frac{1}{2}$$

To solve this,

$$e^{-\lambda M} = \frac{1}{2}$$

$$e^{\lambda M} = 2$$

$$\lambda M = \ln 2 \text{ (taking natural logarithms)}$$

$$M = \frac{1}{\lambda} \ln 2$$

Practical Follow-up 14.2

Compare the mean and variance for your survey results in Section 14.2 with the theoretical values $1/\lambda$ and $1/\lambda^2$.

χ^2 Tests for Our Results

First Survey: Comparison with a Poisson Model

Tables 14.5 and 14.6 give the results of the first survey.

Table 14.5   Time intervals between successive cars arriving at a petrol station

9 s	44 s	7 s	1 min 21 s
51 s	4 min 17 s	2 s	2 min 22 s
3 min 34 s	1 min 31 s	5 min 27 s	27 s
1 min 42 s	5 s	7 s	35 s
28 s	42 s	2 min 1 s	3 s
37 s	30 s	29 s	54 s
3 min 41 s	59 s	11 s	1 min 37 s
48 s	13 s	23 s	4 s
38 s	21 s	38 s	16 s
12 s	51 s	28 s	47 s
1 min 48 s	1 min 12 s	46 s	2 s
51 s	1 min 10 s	29 s	1 min 43 s
2 min 19 s	41 s	31 s	34 s
1 min 10 s	2 min 8 s	4 min 20 s	2 min 14 s
14 s	3 s	2 min 19 s	12 s
1 min 44 s	58 s	30 s	1 min 11 s
1 min 9 s	1 min 6 s	9 s	27 s
13 s	1 min 35 s	29 s	1 min 27 s
2 s	1 min 32 s	39 s	46 s
1 min 24 s	42 s	5 s	1 min
25 s	1 s	2 min 3 s	21 s
8 s	1 min 21 s	50 s	

Table 14.6   Number of cars arriving in each 1 min interval

x	Observed frequency	Expected frequency	$O - E$	$(O - E)^2$	$(O - E)^2/E$
0	29	32.4	−3.4	11.56	0.357
1	38	32.4	5.6	31.36	0.96
2	15	16.2	−1.2	1.44	0.089
3	6	5.4	−1	1	0.143
4 or more	0	1.6			
				Total	$\chi^2 = 1.549$

$v = 2$ (4 classes − 2 restrictions)

The significant value for $\chi_2{}^2$ at 5% level is 5.99. Therefore, our result is not significant and the data are not significantly different from a Poisson distribution $\mu = 1$.

Second Survey: Comparison with an Exponential Model

Table 14.7 gives the results of the second survey.

$v = 5$ (7 classes − 2 restrictions)

The significant value for $\chi_5{}^2$ at the 5% level is 11.07. So our data do not differ significantly from an exponential distribution with $\lambda = 1/61$.

Table 14.7 Time intervals between cars

x	Observed frequency	Expected frequency	$O - E$	$(O - E)^2$	$(O - E)^2/E$
0–	32	33.8	1.8	3.24	0.096
30–	24	25.1	−1.1	1.21	0.048
60–	10	12.6	−2.6	6.76	0.537
90–	9	7.7	1.3	1.69	0.219
120–	6	4.7	1.3	1.69	0.360
150–	0	3.0	−3	9	3.000
180–	1	1.7	1.8	3.24	1.906
210–	2	1.0			
240–	2	0.7			
270–	0	0.3			
300–	1	0.3			
330–	0	0.2			
				Total	6.166

Chapter 15 Simulations

Introduction

A simulation is a model of a real situation. In the same way that a scale model of a new car or plane is built and tested in a wind tunnel, so a mathematical model may be used to test what might happen in a medical or social situation, where experimentation with human subjects might be unethical or too lengthy.

Here are some examples where mathematical models might be used.

1. In studying the spread of disease among human populations or pests among agricultural crops, a simulation might be used to answer a question such as how will the rate of infection increase (for a disease such as whooping cough) if lower percentages of babies are inoculated.
2. In studying the survival of threatened animal species a simulation could be employed to find out, for instance, how low do the numbers in a population have to fall for that group to be threatened by total extinction.
3. In studying queues, we may try to answer such questions as the following. How many pumps are needed at a petrol station? How many pay booths are required at a toll bridge? Is a central queueing system more efficient than queues at separate counters in a bank or Post Office? Will the introduction of an appointments system mean that a doctor's patients will have to wait shorter lengths of time for consultations?

Section 15.1 Steps Involved in Setting Up a Simulation

1. Observe the real situation and identify the important variables. For example, these may be the birth and death rates of a threatened species or the time intervals between the arrivals of patients at a hospital casualty department.
2. Decide on the main processes in the situation. (A flow diagram may be helpful.)
3. Can any of the features be modelled by a mathematical or statistical distribution?
4. What are the assumptions made by these mathematical models? For example, can we assume that casualty patients arrive independently of each other? This assumption would not be true of road accident victims. Check all your assumptions against reality. Are they near enough true?
5. What parameters do you need to know? Will you need to estimate the mean and variance of doctors' consultation times, say, by taking a sample of observations?
6. Take a sample of observations in order to estimate the values of the parameters.
7. Set up your simulation and run it several times.
8. If possible, compare your results with reality.
9. If your simulation is not producing realistic results, try to incorporate additional factors which may be influencing the situation, or try to replace a simplistic relationship between variables with a more sophisticated one.
10. If your simulation is realistic, use it to predict the outcome of changing a feature in the real situation.

Setting up a simulation can in itself help to give a greater understanding of the real situation as we need carefully to consider the important factors and their interaction. By running the simulation, we may see the relative importance of the features involved. If we can test the predictions produced by our model against what happens in reality over a period of time, we can ensure that we have not overlooked any important factors in the simulation. Then the simulation can be used to predict what might happen if certain features were to be changed. For example, if an extra petrol pump was installed in a garage, would it be cost effective?

A very simple simulation will use a deterministic relationship to predict the outcome of a situation. If this does not produce a realistic prediction, the relationship between variables in the model can be made more sophisticated by the introduction of random elements. Here is an example.

A charter company operating cut-price flights to the USA know that some of their passengers do not turn up for the flight they have booked, owing to illness or for some other reason. As profit margins are very small, it would be helpful to predict just how many passengers will fail to check in for a flight.

Model 1 A deterministic model might suggest that 90% of all passengers always turn up. Using this model, the company will book 111 passengers for every 100 seats, but of course on some occasions their prediction will be wrong and they will find themselves overbooked. Unfortunately a simple deterministic model is not very realistic.

Model 2 A random element can be introduced by arguing that each individual passenger has an independent probability of 0.9 of actually checking in to the flight. If this model is run as a simulation, using random numbers or a computer, a different answer will be obtained each time that it is run.

The model being used here is a binomial distribution with $p = 0.9$ and it assumes that passenger non-arrivals are random and independent. As these assumptions are unrealistic, the model will not produce very reliable predictions.

Model 3 The model can be refined still further, by pointing out that 50% (say) of all seats are booked by couples, who will either both travel together, or both stay at home if either one of them is ill (Figure 15.1). It is possible to introduce two stages of randomisation into this model.

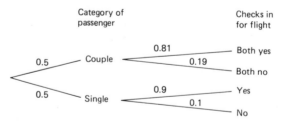

Figure 15.1 Model in which single passengers and couples are considered

Model 4 Finally, a still more sophisticated model might first try to establish the distribution of group sizes of passengers travelling together (Figure 15.2). The probability for each size of party, 1, 2, 3, 4, etc., people, could be found and then the conditional probability that parties of each size party all check in to travel.

The simple model, model 1, would be called a *deterministic* model. It can only generate one answer in a given situation. Models 2, 3 and 4 introduce random elements and are known as *stochastic* processes. This will mean that each of these models will generate a

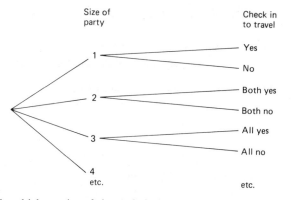

Figure 15.2 Model in which parties of size 1, 2, 3, 4, etc., are considered

different result each time that the simulation is run. The differences between these models can be assessed by running them all with the same set of random numbers and comparing the results.

Section 15.2 Simple Simulations

Simulation 15.1 Collecting Free Gifts

Small plastic models are given away as free gifts with a certain brand of breakfast cereal.

Problem: If there are six different models, how many packets of cereal must I buy on average to collect a complete set?

Simulation Using a Die

Throw a die and record your score at each throw. Continue until you have 'collected' each of the numbers from 1 to 6.

Simulation Using Single-digit Random Numbers

Single-digit random numbers may be used instead of a die, in which case the numbers 7, 8, 9 and 0 should be ignored.

How many packets of cereal did you need to buy in order to collect a complete set of models?

Repeat the entire experiment several times aiming for a total of 50–100 results altogether among the members of your class. What is the mean number of packets of cereal needed to collect a set?

The mathematical model which we have used assumes that the distribution of plastic models among cereal packets is entirely random and that each type has an independent probability of 1/6 of occurring in any packet. Do you think these assumptions are realistic?

A program listing in BBC BASIC for this simulation is included in Section 15.7.

Computer Simulation

A larger version of this problem can be considered if a microcomputer is used.

Picture cards are frequently given as a free gift with a certain brand of tea. If there are 48 in a set, how many packets of tea will I need to buy, on average, in order to collect them all?

Simulation 15.2   Birthdays

Problem: What is the probability that at least two people in any group of 30 will have the same birthday, i.e. day and month, and not necessarily the same year?

You may have already considered this problem mathematically as it appears in Chapter 2, in which case you can compare your simulation results with your calculated probability. A variation of this problem is to find the size of the smallest group for which the probability is $1/2$.

There are several methods of finding a solution to the problem.

1. By calculating probabilities.
2. By obtaining a sample of several groups of 30 people and finding an empirical estimate of the probability. Class registers could be used to do this.
3. By simulation.

Using Dice

1. Throw 2 dice to decide on the day. Allocation of scores as in Figure 15.3 is suggested. Ignore any scores between 6, 2 and 6, 6.

Score on first die

	1	2	3	4	5	6
1	1	2	3	4	5	6
2	7	8	9	10	11	12
3	13	14	15	16	17	18
4	19	20	21	22	23	24
5	25	26	27	28	29	30
6	31					

Score on second die

Figure 15.3   Allocation of scores to decide the date

2. Throw the 2 dice again to decide the month (Figure 15.4). Discard any dates which do not exist such as 31 April.

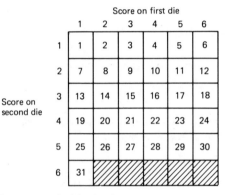

Score on first die

	Odd	Even
1	January	July
2	February	August
3	March	September
4	April	October
5	May	November
6	June	December

Score on second die

Figure 15.4   Allocation of scores to decide the month

3. Record each 'birthday' that you have obtained and repeat the process, stopping either if you get 2 birthdays on the same date or when you reach 30.
4. Repeat the whole experiment several times and then combine the results for your class.
    Calculate an estimate of the probability that at least 2 people in a group of 30 have the same birthday:

$$p = \frac{\text{number of groups containing a repeated birthday}}{\text{total number of trials}}$$

A variation on this problem asks what is the smallest group size for which the probability that at least 2 of the people have the same birthday is at least $1/2$?

Using Random Numbers

There are various ways in which random numbers may be used to select birthdays. Here are two suggestions.

Method A

Allocate each day in the year a number 1–365. Use three-figure random numbers to select dates.

Random numbers 001–365 give the date directly, and another set may also be used, say 401–765, if 400 is subtracted. Any random numbers outside these ranges, e.g. 832 or 384, are discarded.

Method B

1. A two-figure random number is used to select the day. Numbers 01–93 may be used. Divide your random number by 31 and record the *remainder* as the date.
    If the remainder is zero, the date is 31. For example, for $17 \rightarrow 17/31 = 0$, remainder 17, the date is 17; for $85 \rightarrow 85/31 = 2$, remainder 23, the date is 23.
2. Use a two-figure random number to select the month. Numbers 01–96 may be used. Divide your random number by 12 and record the remainder as the month. If the remainder is zero, record the month as December.

Our simulation model assumes that the distribution of births throughout the year is random and uniform. A survey reported in *The Guardian* on 20 March 1984 showed that this is in fact not so and that for the population of Great Britain as a whole there is a peak in births during the spring months (but with variations among the different social classes).

A Further Variant on the 'Birthday Problem'

If people enter a room one at a time, how many people must come in to the room before two have the same birthday? There is a BBC BASIC program listing in Section 15.7 for this version of the simulation.

Section 15.3 Population Models

Simulation 15.3 **Birds**

Problem: **If a pair of birds nest and mate, how many of their offspring (and of the original pair) are likely to survive to the following year?**

This simulation follows through a breeding cycle of 1 year for a pair of non-migrating native British birds. A female will lay a clutch of 10 eggs on average. Of these, not all will be hatched, and the probability that an egg produces a chick surviving to 1 month old is 0.84. If this stage is reached, the probability that the chick survives to 3 months is 0.7. At this stage the fledgeling has to learn to fly and fend for itself; however, the main causes of mortality are the shortage of food and cold weather during the winter months. While half of the adult birds survive the winter, only one-tenth of the 3 month old fledgelings will be alive the following year. The data for this simulation refer to the Great Tit and were given by Begon and Mortimer (1981) (the full reference is given at the end of this chapter).

Running the Simulation

The number of eggs laid by each pair of birds is modelled by a normal distribution with mean 10 and variance 4. The calculations involved in dividing up the area under the normal distribution curve and the allocation of random numbers is shown at the end of this section. A recording sheet for this simulation is provided among the photocopiable pages (page PC14).

Stage 1   Details are given in the instruction sheet of how to use a three-figure random number to determine the number of eggs laid.

Stage 2   For *each egg*, then select a two-figure random number to determine whether the egg hatches. Random numbers 00–83 denote a successful hatching while numbers 84–99 show that an egg does not hatch.

Stage 3   For the next two stages, a single-figure random number only is required for each chick. Stage 3 determines whether the chick survives to 3 months (numbers 0–6) while numbers 7, 8 and 9 denote a mortality.

Stage 4   Stage 4 decides on survival through to the following spring. Of the chicks which live to 3 months, only those allocated the random number 0 will survive to the next year.

Stage 5   To determine whether a parent bird survives, toss a coin for each. Heads (H) denotes a survival and tails (T) mortality.

At the end of the cycle, see how many birds have survived into the following year. Collect up the results for the whole class and compare your findings.

Instruction sheet

Simulation 15.3

Stage 1   Number of eggs laid. Use a three-figure random number (Table 15.1).

Stage 2   Number of eggs hatched. Use a two-figure random number for each egg, i.e. 00–83 means that egg hatches, and 84–99 means that egg does not hatch. (Alternatively, throw a die once for each egg. Score 1, 2, 3, 4 or 5 means that egg hatches, and score 6 means that it does not hatch.)

Stage 3   Number of chicks surviving to 3 months (fledgelings). Use a single figure-random number. Numbers 0–6 mean that chick survives and 7–9 mean that chick does not survive.

Stage 4   Number of fledgelings surviving to next year. Use a single-figure random number. Number 0 means that fledgeling survives and 1–9 mean that it does not survive.

Stage 5   Number of parent birds surviving to next year. Toss coin. Heads means that parent birds survive, and tails means that they do not.

Table 15.1 Random numbers for the number of eggs laid

Random numbers	Number of eggs
000–002	4
003–011	5
012–039	6
040–105	7
106–226	8
227–400	9
401–598	10
599–772	11
773–893	12
894–959	13
960–987	14
988–996	15
997–999	16

Allocation of Random Numbers to Results

For the number of eggs laid, as a model use a normal distribution with $\mu = 10$ and $\sigma = 2$ (Figure 15.5 and Table 15.2).

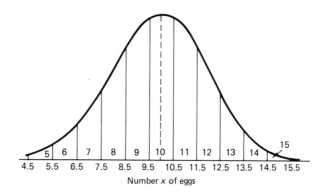

Figure 15.5 Allocation of random numbers to the results

Table 15.2 Allocation of random numbers to the results

x	z	Area $F(z)$ to left	Random numbers
4.5	−2.75	0.003	000–002
5.5	−2.25	0.012	003–011
6.5	−1.75	0.040	012–039
7.5	−1.25	0.106	040–105
8.5	−0.75	0.227	106–226
9.5	−0.25	0.401	227–400
10.5	0.25	0.599	401–598
11.5	0.75	0.773	599–772
12.5	1.25	0.894	773–893
13.5	1.75	0.960	894–959
14.5	2.25	0.988	960–987
15.5	2.75	0.997	988–996
16.5	3.25	1	997–999

Simulation 15.4   **Blue Whales**

The Blue Whale is the largest of the whale species (a few have been found to measure over 30 m long and weigh over 150 tonnes) and as a consequence was one of the most intensively hunted. It is estimated that today about 1000 Blue Whales survive in the southern hemisphere compared with an original stock of 200 000 before modern commercial whaling began.

In 1986 the International Whaling Commission imposed a total ban on all commercial whaling but, despite that, 600 whales were caught in that year by nations refusing to acknowledge the ban.

Like most whales, the female Blue Whale will produce only a single calf once every 2 years. Gestation takes almost 1 year and the calf is fed on its mother's milk for the best part of a further year. As a result, populations increase very slowly, even when left undisturbed.

A Model for the Population Growth of Blue Whales: Survival Rates from One Age Group to the Next

For most age groups in an undisturbed population, the survival rate from one year to the next is estimated to be about 95%. Older whales and new-born calves have lower survival rates. It is estimated that only one-half of all new-born calves survive their first year. If we work in time periods of 2 years (the length of a breeding cycle), we obtain the estimated probabilities given in Table 15.3.

Table 15.3   Estimated probabilities

Age at the beginning of a 2 year cycle (years)	Probability of survival into the next 2 year period (approximate value to one decimal place)
0–1	0.5
2–29	0.9
30 +	0.8

The average lifespan of a Blue Whale is about 30 years but there have been recorded cases of animals over 100 years old.

Breeding Cycles

Breeding does not start until the fifth year (for a minority of females) and full sexual maturity occurs at age 7 years for the majority. Approximately one-half of females in the mature age groups will produce young in any one year, giving a proportion of 90% over a period of 2 years. As breeding for any one female will not occur more often than once every 2 years, it is convenient to work in 2 year periods. The estimated probabilities for breeding are given in Table 15.4.

As there has been little direct observation of whales, most of the available information was gleaned in the past by scientists examining slaughtered whales in the whaling factories.

Table 15.4 The estimated probabilities for breeding

Age of female (years)	Probability of producing a calf in a 2 year period
0–3	0
4–5	0.2
6–7	0.4
8–25	0.9
Over 26	0.5

Sex ratio

The sex ratio of females to males is generally accepted as being approximately 1 to 1.

Age Distribution of a Population of Approximately 1000 Blue Whales

Since the actual age distribution of the whale population is not known, we have estimated approximate numbers of males and females in each 1 year age group by assuming the death rates quoted above and an approximate steady state.

The final frequencies are estimated from the survival rates for each age group and involve some rounding. The total frequency for the table is in fact 527 females (and thus 527 males), giving a combined population of 1054. Frequencies for adjacent age groups can be combined to give frequencies for 2 year age groups.

Smaller Populations

The age distribution for a smaller group can be estimated by dividing by a suitable factor.

Estimated Age Distribution of Approximately 1000 Blue Whales

Figures are given for one sex only (Table 15.5).

Table 15.5 Age distribution

Age (years)	Frequency	Age (years) (continued)	Frequency (continued)
0	58	20	11
1	29	21	10
2	27	22	10
3	26	23	9
4	25	24	9
5	24	25	8
6	22	26	8
7	21	27	7
8	20	28	7
9	19	29	7
10	18	30	5
11	17	31	4
12	16	32	3
13	16	33	3
14	15	34	2
15	14	35	2
16	13	36	1
17	13	37	1
18	12	38	1
19	12	39	1
		40	1

Total 527 of each sex.

Running the Simulation

Firstly, decide on the size of your population group. We decided to scale down the size of the simulation to a group of just over 100 whales, to see how quickly they would increase their numbers or indeed whether such a small number would be in danger of extinction.

The frequencies for *2 year* age groups for each sex are given on the instruction sheet. These frequencies were calculated from the previous set for approximately 1000 whales by dividing each frequency by 10 and adding consecutive classes to give 2 year age groups.

Class Organisation

It is easier to work in small groups of pupils and, in a large class, you may have several groups working independently. Half of each group should work on a recording sheet for the female whales (page PC15) and the other half on a recording sheet for the males (page PC16). One recording sheet is needed for each 2 year cycle and a simulation should run for at least 5 cycles (10 years).

1. *Transition to the next age group.* Allocate each whale in your starting population a single-figure random number to determine their survival into the next 2 year cycle. Record all your random numbers on a copy of the appropriate recording sheet included in the photocopiable pages (PC15 and PC16). Separate (and *different*) sheets are used for males and females. The instruction sheet gives details of which random numbers indicate survival. Survivors can later be transferred to a new sheet for the next 2 year cycle. Remember, however, to write them in as the next age group going down the sheet.

2. *Breeding* (female population only). Allocate each of the females who are going to survive a further random number. The outcome of this allocation determines whether that female produces a calf during that 2 year period. Again refer to the instruction sheet for details.

    To simplify the situation, random numbers are not allocated to females who die during the 2 year cycle as any unborn or suckling calf would certainly die with the mother.

3. *Sex of calves.* For each calf born, toss a coin to decide its sex. Then add the new calves onto the new sheet (for the correct sex) as the 0–1 year age group for the next 2 year period. There are two specimen recording sheets (Tables 15.9 and 15.10) showing how they should be filled in, at the end of this section.

    At the end of 5 cycles, compare the size of your population with the size of the group that you started with. Has the population increased or decreased in size?

    If you can run this simulation on a microcomputer, try to find the largest size of a group which would be threatened by extinction.

We are grateful to Dr Justin Cooke, for his help in devising this simulation.

Instruction sheet

Simulation **15.4**

Starting population

Frequencies are given for one sex only (Table 15.6).

Table 15.6 Frequency distribution

Age (years)	Frequency
0–1	9
2–3	5
4–5	5
6–7	4
8–9	4
10–11	4
12–13	3
14–15	3
16–17	3
18–19	2
20–21	2
22–23	2
24–25	2
26–27	2
28–29	1
30–31	1
32–33	1
Total	53

1. *Transition to next age group* (Table 15.7)

Table 15.7

Age at beginning of 2 year period (years)	Random numbers	
	Live	Die
0–1	0 1 2 3 4	5 6 7 8 9
2–29	0 1 2 3 4 5 6 7 8	9
30+	0 1 2 3 4 5 6 7	8 9

2. *Breeding* (Table 15.8)

Table 15.8 Breeding

Age (years)	Random numbers	
	1 calf born	No calf born
0–3	—	—
4–5	0 1	2 3 4 5 6 7 8 9
6–7	0 1 2 3	4 5 6 7 8 9
8–25	0 1 2 3 4 5 6 7 8	9
26–29	0 1 2 3 4	5 6 7 8 9

3. *New-born calves*
   Toss a coin for male or female.

Table 15.9 Specimen sheet 1 for Simulation 15.4 (Blue Whales; females; 2 year cycle number 1)

Age (years)	Numbers at start of cycle	Random numbers for survival	Number of survivors	Random numbers for breeding	Number of calves
0–1	9	7 7 8 0 3 6 1 8 7	3		
2–3	5	2 8 6 2 4	5		
4–5	5	5 6 8 2 2	5	4 9 6 7 3	—
6–7	4	0 4 5 2	4	3 5 5 2	2
8–9	4	6 7 0 1	4	5 1 2 1	4
10–11	4	3 7 9 1	3	5 4 5	3
12–13	3	1 3 0	4	1 1 8 3	4
14–15	3	0 1 6	3	2 7 0	3
16–17	3	5 8 2	3	9 7 7	2
18–19	2	4 0	2	2 9	1
20–21	2	4 4	2	9 2	1
22–23	2	6 7	2	9 5	1
24–25	2	2 8	2	7 2	2
26–27	1	3	2	1 5	1
28–29	1	2	2	4 2	1
30–31	1	8	0		
				Total	25

Number of female calves, 11.
Number of male calves, 14.
(Add these numbers to the appropriate sheet for the next cycle as the 0–2 years age group.)

Table 15.10 Specimen recording sheet 2 for Simulation 15.4 (Blue Whales; females; 2 year cycle number 2)

Age (years)	Numbers at start of cycle	Random numbers for survival	Number of survivors	Random numbers for breeding	Number of calves
0–1	11	4 1 2 6 6 9 9 4 3 3 1 7			
2–3	3	6 5 7	3		
4–5	5	9 2 6 6 3	4	9 7 1 6	1
6–7	5	6 6 4 7 6	5	1 7 1 3 3	4
8–9	4	8 8 1 4	4	6 5 0 4	4
10–11	4	2 7 9 4	3	9 0 2	2
12–13	3	2 9 4	2	8 2	2
14–15	3	9 8 8	2	0 2	2
16–17	3	6 3 8	3	5 3 7	3
18–19	3	6 6 3	3	7 9 1	2
20–21	2	0 5	2	6 9	1
22–23	2	3 9	1	1	1
24–25	2	2 5	2	4 9	1
26–27	2	2 2	2	4 1	2
28–29	1	3	2	7 0	2
30–31	1	8	1		
				Total	25

Number of female calves, 14.
Number of male calves, 11.
(Add these numbers to the appropriate sheet for the next cycle as the 0–2 years age group.)

Section 15.4   Queueing Simulations

Simulation 15.5   Road Junction

Figure 15.6 represents a one-way traffic lane on a dual-carriageway main road, joined by a slip road or feeder lane.

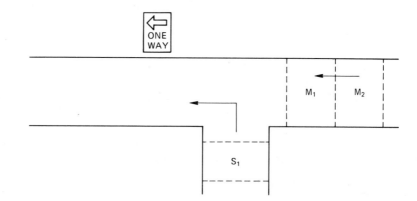

Figure 15.6   One-way traffic lane on a dual-carriageway main road, joined by a slip road or feeder lane

In each time period of 10 s a vehicle approaches from direction M along the main road with probability $P(M)$ and a vehicle approaches from the slip road with independent probability $P(S)$. Vehicles at S can only join the main road if there are two successive gaps in the main road traffic; otherwise they form a queue.

Problem: **What is the average length of the queue at S? Would installing traffic lights help?**

Running the Simulation

It may be easiest to use a copy of the diagram with counters to represent vehicles. Work through the following steps for each 10 s time period, using Recording sheet 1 on photocopiable page PC17.

1. If there is a counter at M_1, let it move on past the junction.
2. If there is a counter at M_2, move it on to M_1.
3. Toss a coin. If the result is heads, place a counter on M_2 to show a newly arrived vehicle from direction M.
4. Throw a die. If the result is 5 or 6, place a counter on square S_1 to show that a car has arrived from direction S. If cars cannot move onto the main road, a queue will form here.
5. If M_2 and M_1 are both empty, allow one car from S_1 to leave the queue and to proceed along the main road.
6. Record all the events for each 10 s time period on the recording sheet provided, using a tick in columns 1–4 and a number in column 5. If there are no vehicles waiting at S, record the queue length as 0.

Repeat the simulation for 100 time periods. Find the total number of vehicles passing from M and S and the average length of the queue at S.

Traffic Lights

All cars arriving at the junction must form a queue in the relevant square if the traffic

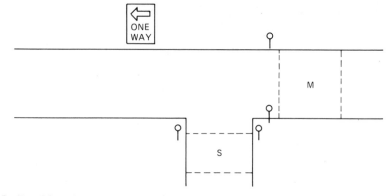

Figure 15.7   Road junction with traffic lights

lights are red (Figure 15.7). Note that we only need one square at M as vehicles arriving at S are no longer waiting for two successive gaps in the main road traffic.

Traffic Light Settings

These may be altered to give the minimum queue lengths (queues will now occur at M on the main road as well as at S on the slip road).

   Here is a suggested pattern to start with. Allow the main road traffic to move for 6 periods of 10 s, and then change the signal to allow slip road traffic to move for 4 periods of 10 s. In any 10 s period, up to 2 vehicles can move through the lights if they are showing green. (Continue to throw the die and to toss the coin to determine arrivals from S and M.)

   Use Recording sheet 2 to record your results. This is included among the photocopiable pages as page PC18.

Section **15.5   Using Random Numbers to Simulate the Selection of a Sample from a Probability Distribution**

Sampling from a Discrete Distribution

Random numbers may be used to select a sample from any discrete distribution. The binomial distribution with $n = 4$, $p = 0.5$, is used as an example. The probabilities corresponding to each value of x are given in Table 15.11.

Table 15.11   Probabilities for $B(4, 1/2)$

x	$P(X = x)$	$P(X \leqslant x)$
0	0.0625	0.0625
1	0.25	0.3125
2	0.375	0.6875
3	0.25	0.9375
4	0.0625	1.0000

If we wish to sample from $B(4, 1/2)$, then the probability that $x = 0$ is selected must be 0.0625 (at each selection) and the probability that we include $x = 1$ at any selection must be 0.25 and so on.

The final column in the table gives the distribution function, i.e. the cumulative probabilities from the second column. If we take a random number y between 0 and 1, we can use it as shown in Table 15.12 to select a value of x from the distribution.

Table 15.12 Selection of x based on a random number y

$$\left.\begin{array}{l} 0.0000 < y \leqslant 0.0625 \\ 0.0625 < y \leqslant 0.3125 \\ 0.3125 < y \leqslant 0.6875 \\ 0.6875 < y \leqslant 0.9375 \\ 0.9375 < y \leqslant 1 \end{array}\right\} \text{ put } \left\{\begin{array}{l} x = 0 \\ x = 1 \\ x = 2 \\ x = 3 \\ x = 4 \end{array}\right.$$

This will give the various values of x with the appropriate probabilities and we get a valid simulation.

If we draw a graph of the distribution function (Figure 15.8), we can see its connection with the procedure outlined above.

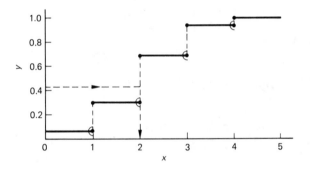

Figure 15.8 Distribution function of y

The y axis can be seen as the random numbers between 0 and 1, and by reading across and down we arrive at the corresponding x value. The diagram shows how a random number $y = 0.4200$ leads to $x = 2$.

This technique can be used to simulate the selection of a random sample from any distribution with a discrete probability function.

Sampling from a Continuous Distribution

The method used above can be adapted to simulate sampling from a continuous distribution. In this case the distribution function will be a smooth curve and we can read directly the appropriate x value for a given y (random number (Figure 15.9).

In most cases, we know the probability density function $f(x)$, and the distribution function $F(x)$ can be found from it by integration (Figure 15.10). For example, if we take the continuous uniform distribution over the range $0 \leqslant x \leqslant 2$,

$$f(x) = 1/2$$

$$F(x_i) = P(X \leqslant x_i)$$

$$= \int_0^{x_i} \frac{1}{2} \, dx$$

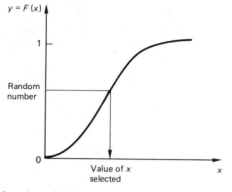

Figure 15.9 Distribution function of y for a continuous distribution

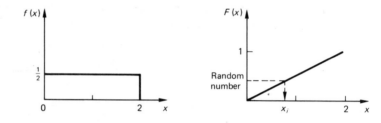

Figure 15.10 Probability distribution function $f(x)$ and distribution function $F(x)$ for $U(0, 2)$

$$= \left[\frac{x}{2} \right]_0^{x_i}$$

$$= \frac{x_i}{2}$$

Sampling from the Normal Distribution

Statistical tables give the distribution function $F(z)$ for the standard normal distribution $N(0, 1)$; so this makes sampling from the normal distribution relatively straightforward. An example is shown in Section 15.6.

Section 15.6   A More Complex Queueing Simulation

Simulation 15.6   Petrol Station

Problem: Are more pumps needed?

Our local self-service petrol station has 4 pumps at present but the manageress wishes to investigate whether more are needed. In order to simulate the effect on waiting times of 1 or more new pumps, we need a model of the present situation. We will need to observe the following features.

1. The time intervals between the arrival of one car and the next.
2. The time spent by each customer buying petrol.

3. The number of cars queueing.
4. The time they spend waiting.

Factors 3 and 4 can in fact be calculated if detailed observations are taken of 1 and 2. A non-rush-hour survey gave the following results.

The average interval between arrivals was just over 1 min and the distribution of these time intervals was very close to an exponential distribution. The exponential distribution does seem to be a convenient model for arrival times, whether for cars arriving at a garage, customers arriving at a bank or shop, or telephone calls coming into a switchboard. (Chapter 14 gives a full explanation.)

The average time spent by customers buying petrol was found to be 3 min 20 s with a standard deviation of 50 s (both measures rounded to the nearest 5 s). The observed times were close to those generated by a normal distribution with the same mean and standard deviation.

Running the Simulation

For each vehicle arriving at the petrol station a four-figure random number is used and taken as a four-figure decimal to represent the cumulative probability (i.e. the area under the normal distribution curve) to calculate the time spent buying petrol.

A second four-figure random number is used to calculate the time interval between the arrival of this car and the next one. Again the random number is used as a four-figure decimal to represent the cumulative probability, this time under the exponential distribution curve.

A Worked Example: Time taken to buy petrol

The model used is a normal distribution with a mean of 3 min 20 s and a standard deviation σ of 50 s. The four-figure random number is 6829. The area $\Phi(z)$ under the curve is 0.6829 (Figure 15.11).

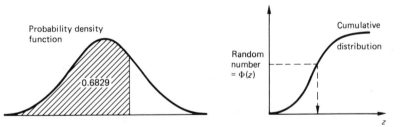

Figure 15.11 Worked example: probability density function and cumulative distribution for the normal distribution

1. Find z values from tables. Here, for $\Phi(z) = 0.6829$, $z = 0.476$.
2. Calculate x value:

$x = $ mean $+ (z \times$ standard deviation)

$= 3$ min 20 s $+ (0.476 \times 50$ s$)$

$= 3$ min 20 s $+ 23.8$ s

$= 3$ min 44 s (to nearest second)

Time interval to Next Arrival

The model is an exponential distribution with a mean of 1 min. The distribution function

$F(t) = 1 - e^{-t}$ and the four-figure random number is 2668:

$F(t) = 0.2668$

The probability density function is $f(t) = e^{-t}$ (Figure 15.12).

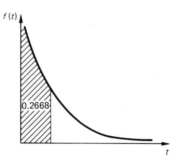

Figure 15.12 The probability density function $f(t) = e^{-t}$

To calculate the time t in minutes,

$F(t) = 0.2668$

$1 - e^{-t} = 0.2668$

$e^{-t} = 1 - 0.2668$

$\quad = 0.7332$

$e^t = \dfrac{1}{0.7732}$

$\quad = 1.3639$

$t = \ln(1.3639)$

$\quad = 0.31 \text{ min}$

$\quad = 18.6 \text{ s}$

So $t = 19$ s to the nearest second.

The steps required to calculate t are as follows.

1. Write a four-figure random number as a four-figure decimal ($\rightarrow 1 - e^{-t}$).
2. Subtract it from 1 ($\rightarrow e^{-t}$).
3. Take the inverse ($\rightarrow e^{t}$).
4. Take the natural logarithm ($\rightarrow t$).
5. Multiply by 60 to get time in seconds.

Recording Sheets

Recording sheet 1 is used to record the random numbers, the times taken for buying petrol and the intervals between arrivals. Recording sheet 2 is used for calculating the time elapsed since the beginning of the simulation for all arrivals and departures. These recording sheets are to be found in the section of photocopiable pages (the recording sheets are pages PC19 and PC20). Tables 15.13 and 15.14 are two specimen sheets filled in for a short demonstration run of 10 cars. Run the simulation for yourself for 50–100 cars.

Table 15.13 Specimen recording sheet 1 for Simulation 15.6: demonstration run for 10 cars

	Time interval since arrival of previous car			Time taken to buy petrol		
Car number	Random number	Time		Random number	z	Time
1				6829	0.476	3 min 44 s
2	2668	19 s		5531	0.133	3 min 27 s
3	3314	24 s		0264	−1.36	2 min 12 s
4	7112	75 s (1 min 15 s)		2102	−0.806	2 min 40 s
5	2986	21 s		8854	1.22	4 min 21 s
6	9596	193 s (3 min 13 s)		8948	1.252	4 min 23 s
7	6514	63 s (1 min 3 s)		1202	−1.174	2 min 21 s
8	9450	174 s (2 min 54 s)		3564	−0.368	3 min 02 s
9	5843	53 s		9207	1.41	4 min 30 s
10	7408	81 s (1 min 21 s)		5208	0.052	3 min 23 s

Table 15.14 Specimen recording sheet 2 for Simulation 15.6: demonstration run for 10 cars

Time from start	Arrivals	Departures	Number of pumps vacant	Cars in queue and waiting time
0 s	1		3	
19 s	2		2	
43 s	3		1	
118 s (1 min 58 s)	4		0	
139 s (2 min 19 s)	5		0	1 (36 s wait)
2 min 55 s		3	0	
3 min 44 s		1	1	
3 min 46 s		2	2	
4 min 38 s		4	3	
5 min 32 s	6		2	
6 min 35 s	7		1	
6 min 40 s		5	2	
8 min 56 s		7	3	
9 min 28 s	8		2	
9 min 55 s		6	3	
10 min 49 s	9		2	
11 min 42 s		9	3	
12 min 10 s	10		2	
12 min 32 s		8	3	
13 min 31 s		10	4	

1. What is the longest queue formed?
2. What is the mean queue length?
3. What is the longest waiting time?
4. What is the mean waiting time?
5. How many pumps would be needed to ensure that no-one had to wait?

You can then change the simulation to see what happens in the following circumstances.

a. 1 pump is out of order and only 3 are available for use.
b. Cars arrive more frequently at rush-hour times—on average one every 45 s.

You may like to use a simple diagram to represent the petrol station with four pumps, and to use counters to represent cars arriving and departing.

Section **15.7** **Random Numbers and Computers**

It is possible on all modern microcomputers to generate a sequence of numbers which can be taken as random. They are not genuinely random since they are generated from an algorithm and hence the sequence is pre-determined once a starting point has been designated. Strictly, they are pseudo-random numbers which, to a greater or lesser extent, have all the properties of a random-number sequence. For important work, it may be necessary to investigate the random number generator mechanism to make sure that it is not putting unwanted patterns into a simulation. For our purposes, we shall assume that the numbers can be taken as random.

The actual language used will differ from computer to computer and even within different dialects of, say, BASIC. It should be possible to do each of the following.

1. Generate a sequence of numbers randomly from the integers 1 to n to simulate, for example, throws of a dice when $n = 6$.
2. Generate a sequence of numbers randomly from the range from 0 to 1 on a uniform distribution.
3. Set the random-number generator so that the same sequence of random numbers can be generated.

In BBC BASIC, RND(n) does item 1, RND(1) does item 2 and RND($-X$) will set the random-number generator to repeat a sequence.

Examples of Some BASIC Programs

Simulating the Collection of 6 Plastic Toys from Packets of Breakfast Cereal

The basic problem is that $p = 1/6$ for each of the 6 toys which may occur independently and we need to check whether we already have this toy and count how many toys before we have one of each.

RND(6) will tell us which toy we have. Stores from T(1) to T(6) are used to indicate whether we have received that toy: 0 = NO; 1 = YES; Count = number of toys received so far; T = number of different toys received so far. So a simple program for one simulation, in BBC BASIC is

```
10  DIM T(6)

100  Count = 0

110  T = 0

120  REPEAT

130  R = RND(6)          Get a toy

140  Count = Count + 1   Add 1 to the count

150  IF T(R) = 0 T = T + 1   If not already collected add 1 to the number of different toys

160  T(R) = 1            Note that this one is collected

170  UNTIL T = 6         Repeat until all toys collected

200  Print Count         Print how many toys collected
```

If we want to repeat the simulation 100 times, we can do so by putting the main program into a loop, but we have to reset all the T(R) stores to zero before starting each simulation.

Call lines 100–170 above, the procedure for simulation, PROCsim. The simulation can then be repeated 100 times by using

```
10 DIM T(6)
2 FOR I = 1 to 100
30 FOR J = 1 to 6
40 T(J) = 0
50 NEXT J
60 PROCsim
70 PRINT Count
80 NEXT I
90 DEFPROCsim
180 ENDPROC
```

This will print out a sequence of 100 numbers showing how many cards were collected each time. If we want to find the frequency distribution, we can store each result and then print out later.

Keep the results in C(Count). Theoretically, this could be infinite but is very unlikely to be greater than 50. To avoid the program crashing, we put Count = 50 if it is greater than 50. Change the above program to have

```
10 DIM T(6), C(50)
70 IF Count > 50 THEN Count = 50
75 C(Count) = C(Count)+1
85 PROCprint:END
```

with the Procedure PROCprint defined as

```
200 DEFPROCprint
205 VDU2
210 PRINT "Number of Toys", "Frequency"
220 FOR I = 1 TO 50
230 PRINT I,C(I)
240 NEXT I
245 VDU3
250 ENDPROC
```

Lines 205 and 245 turn a printer on and off. If you do not have a printer, omit these lines and change line 220 to FOR I = 6 to 28. This gives a screenful of data showing how many times the collection was made in $6, 7, \ldots, 28$ cereal packets. Clearly there must be at least 6; the number of times that it took more than 28 packets can be found by subtracting the total from 100.

The Birthday Problem: Variant Version

Simulation of how many people enter a room before any two have the same birthday can be done as follows.

We make the assumption of 365 days in a year and births equally likely on each day.

We use D(I) to keep records of which of the 365 days have been birthdays; N(I) keeps track of how many people are in the room when the first repeat birthday occurs.

```
10  DIM D(365),N(200)

100  DEFPROCsim

110  Count=0:Cycle=0

120  REPEAT

130  Count=Count+1

140  P=RND(365)

150  IF D(P)=1 Cycle=1

160  D(P)=1

170  UNTIL Cycle=1

180  N(Count)=N(Count)+1

190  ENDPROC
```

To do the simulation once, or just to print out the sequence of results as they happen, replace 180 to read 180 PRINT Count.

Reference

Begon, M., and Mortimer, M., 1981, *Population Ecology—A Unified Study of Animal and Plants*, Blackwell Scientific, Oxford.

Chapter 16 Practical Investigations and Longer Projects

This chapter differs from the other chapters in this book. There are no practicals for you to do. It is about the general principles of carrying out a statistical project.

Forward Planning

This is vital. By thinking through the stages of your project before you start, you can anticipate many problems and decide whether your intended investigation is in fact viable. It is better to abandon an impractical project at this stage rather than later on after much time and energy have been invested in it.

Finding a Topic Which Interests You

You may find actually getting started on a project the most difficult hurdle to overcome. If so, the following ideas may give you some inspiration.

1. Consider your own hobbies and interests.
2. Use current topics of interest in the local or national news.
3. Many of the short practicals outlined in the chapters of this book can be expanded into larger investigations.
4. You need not necessarily collect your own data. There may be many sources of secondary data in your locality. These include the local newspaper, the local council, weather-recording stations and so on.

Formulating the Question

At this stage, it is important that you have a clear question in mind. Exactly what is it that you want to find out?

Having located a topic of interest, read as much published information as you can find to help you to formulate an interesting (and possibly useful) problem. Then think about your problem in the following terms. Are you trying to describe a situation? Have you chosen an area for which there is little information available, and which can be usefully investigated on an exploratory level? Alternatively, at a higher level, you may be trying to compare two situations or groups. At a higher level still, you may be trying to test a hypothesis which perhaps can be generalised to other situations.

Here are some examples of formulating questions.

50 years ago, many people left school at 14, started work and in many respects were treated as adults. Today, everyone remains at school until 16, 18 or even 19 and are dependent on their parents for much longer. Is childhood being extended and adulthood deferred? What precise questions could we investigate? Some ideas are suggested here as follows.

1. Do teenagers today have fewer responsibilities than their parents did?
2. Do teenagers today have less spending power than their parents did?
3. Are teenagers more politically aware than they were?
4. Have higher divorce rates put more pressure on children? Do they have to take a greater share of domestic responsibilities in a single-parent household?

A second example of an area of study might be road junctions and roundabouts. Here are some questions which could be generated from the original idea.

a. Are traffic lights safer than roundabouts?
b. Are mini roundabouts dangerous?
c. Do drivers make more mistakes at roundabouts than at traffic lights?
d. Is there a relationship between the size of a roundabout and the number of mistakes that drivers make in negotiating it?

What Data Do You Need?

Go back to your problem again and decide what information you will need to answer your question. What factors are you concerned with, and how can you measure them?

We shall consider our two sets of examples here to show what information might be needed in order to answer each question.

1. *Do teenagers today have more responsibilities than their parents did?* How can we measure responsibilities? Should we present our sample with a checklist of household chores: cooking meals, caring for younger children, etc.? Should we measure the frequency of each activity or the time spent on each activity? Should we include paid work outside the home? Should we consider or include any other activities?

2. *Do teenagers today have less spending power than their parents did?* In order to investigate this problem, we shall need to overcome the problem of inflation by concentrating on items bought rather than on amounts of money spent.

3. *Are teenagers more politically aware than they were?* Can we measure political awareness? Certainly, people can vote now at 18 years which they could not do 20 or 30 years ago. Were the parental age group members of trade unions at age 15 or 16 years? Are there other indicators of political involvement?

4. *Do teenagers have to take a greater share of domestic responsibilities in a single-parent household?* The variables to be measured here are similar to the first problem. In addition, we shall need to define and identify single-parent households, to include divorced, widowed and separated parents.

Now for our second set of problems.

a. *Are traffic lights safer than roundabouts?* How do we measure 'safeness'? Is the number of accidents per year, say, a measure of lack of safety?

b. *Are mini roundabouts dangerous?* The same type of measure will be needed for problem b as for problem a. Also are we going to compare mini roundabouts with other junctions—if so which ones? How can we be sure that the comparisons made for problems a and b are really valid? Ideally, these surveys would best be conducted at a single road junction before and after a change in road layout is made. Your local council will have details of any such proposed changes and you may well be lucky enough to find that traffic lights are to be introduced at a particular junction or a mini roundabout taken out of service.

c. *Do drivers make more mistakes at roundabouts than at traffic lights?*

d. *Is there a relationship between the size of a roundabout and the number of mistakes that drivers make in negotiating it?* For problems c and d, we must decide which categories of drivers we wish to include: heavy-goods vehicle drivers, bus drivers, motor cyclists and pedal cyclists? Should we measure the size of the roundabout by its internal diameter, its external diameter or the number of roads converging into it? (or a combination of these). Also we need to consult the highway code to determine the correct way of negotiating a roundabout or road junction to go left, right or straight ahead, so that possible mistakes can be listed in advance. When recording our observations, do we need to consider the number of other vehicles present? For instance, if only one car is on the roundabout or junction, does any mistake(s) matter? Are some mistakes more important than others?

Can You Obtain the Data that You Need?

It is at this stage that you should seriously consider the feasibility of your project. Go through the list of measurements that you require and decide realistically whether you will be able to obtain them, and how. If there are going to be too many problems in obtaining the relevant data, you may have to consider changing your investigation. Consider the feasibility of our examples.

1. *Do teenagers today have more responsibilities than their parents did?* Having decided on the age group to be considered here (13–16 or 13–19 years?), we need to decide how to collect the information we need. Should we use a questionnaire or an interview, both of which rely on people remembering their activities over a period of time (say the previous week or month). Or should we provide our sample with a diary to record their tasks over the subsequent week or month? An alternative might be to interview the parents of the sample if we suspect that teenagers may exaggerate the extent of their domestic responsibilities! Also we must take account of the fact that our survey will highlight this area and possibly cause our sample to make more effort for the period of the project.

   Our main problem will be to collect comparable data from parents as to their responsibilities when they were teenagers. Will people's memories be reliable over a period of 15, 20 or 25 years? We certainly cannot hope to gain accurate information from this group and, although grandparents might be called on to substantiate evidence (!), it might be better to consider the possibility of a change at this point.

Similar problems arise when we consider problems 2 and 3.

2. *Do teenagers today have less spending power than their parents did?* Do we just include personal purchases and spending or can we include gifts and presents from parents and grandparents who may buy clothes, bicycles, computers, etc.? When investigating parents' spending power as teenagers, we shall again have to rely on their memories. Will they be inclined to under-state their spending power? Will they tend to forget minor purchases such as magazines, make-up, records, hair-cuts, etc.
3. *Are teenagers more politically aware than they were?* Of all the problems outlined in this section, this one presents the most problems for data collection. The two age groups are not strictly comparable, as the age for voting has been reduced to 18 years. We could develop a questionnaire to test the political knowledge of teenagers today, but a retrospective one for parents is not feasible.
4. *Do teenagers have to take a greater share of domestic responsibilities in a single-parent household?* The best alternative may well be to compare present-day teenagers living in one-parent and two-parent homes. This would overcome problems caused by lapses of memory over long periods of time. However, we would need to be sensitive to the fact that children who have lost one parent, through whatever cause, may be reluctant to take part in a survey.

If we decide to stay with our original idea we may have to be content with obtaining a classification of the types of responsibility undertaken by the older generation, rather than the extent of these responsibilities and the time spent on them.

If a questionnaire is to be used, it should be carefully drawn up and a pilot survey made on a small sample to ensure that the questions are unambiguous and extract the required information. At the end of this chapter, we give some basic guidelines for compiling a questionnaire, but you should consider the following points. Are any of the questions open to misinterpretation? Is the wording clear? Will these questions give you the information you require? Are any of the questions unnecessary?

Now consider our second set of problems a–d. Providing suitable road junctions can be found of the type(s) required, and enough helpers are available for observing and recording data, these projects should present few problems at this stage. All the data required can be gathered by direct measurements and observations.

How Large Will Your Sample Be?

The answer to this question depends to a large extent on the time and the amount of help available to you and the accuracy of the estimate or answer required. However, if you plan to conduct any statistical tests of significance (see a later section), you should aim to collect a sample of 50 to 100 observations in each group for which comparisons are to be made. In exceptional circumstances, significance tests may be conducted on small samples using the Student t distribution, provided that the measurements in the population can be assumed to follow a normal distribution (refer to Chapter 10).

If a significance test is to be used to detect a small difference (in sample means, say), then large samples will be needed to gain a significant result. A pilot survey is useful in this respect as it will indicate the likely magnitude of any differences and the size of the samples can be calculated accordingly.

How Will You Select Your Sample?

If you have not already done so, decide on your target population, i.e. the group(s) about which you need information in order to answer your original question. That decided, you must consider next how you can select a sample or smaller group from that population. If possible, you should try to select a random sample, as then the variability from sample to sample can be predicted mathematically. The sampling distributions of statistics such as sample means, variances and proportions, etc., can be determined, and most standard statistical hypothesis testing assumes that the results have been obtained from a random sample.

To select a random sample, you require a sampling frame or list of the items of the target population so that each member can be numbered and the sample drawn using random numbers (refer to Chapter 6). In many situations, such a list will not exist. Our second problem concerning the behaviour of drivers at roundabouts is an example of a study for which a sampling frame for drivers would not be available (but we can take a random sample of roundabouts in the area, and times of day or days of the week).

An alternative method of sampling may be used if a random sample is not possible, but any statistical results should be regarded with caution. Even if a random sample is used, your results can only be generalised to the restricted population from which your sample is drawn.

Non-random sampling schemes are outlined at the end of this chapter.

How Are You Going to Record Your Results?

Your data collection and recording needs to be carefully planned. If you plan to collect measurements or direct observations, draw up a clear recording sheet with adequate space for all the information needed. This will save much panic and confusion in the heat of the moment. However, bear in mind that, while recording results directly onto a tally chart, say, organises your data collection and can make subsequent analysis easier, it does lose the original detail of the raw data. It is worth spending time at this stage deciding how you plan to analyse the data and how much detail you will require. A stem plot (or stem-and-leaf diagram) is a useful technique for recording data. It preserves much of the detail of the original data but even so gives no record of the order in which the data were collected. (The order matters if time is an important factor.)

If your project involves a questionnaire or interview, you should run a small pilot survey. This will hopefully reveal any problems and give you the opportunity of eliminating these in advance. Also you will be able to see what kind of answers you are likely to obtain and may be able to consider a pre-coded recording sheet which classifies the answers straight away. While this does simplify analysis, it also restricts the range of answers that can be recorded.

What Types of Results or Measurement Will You Obtain?

Some questionnaire and interview surveys may yield a great deal of qualitative information, i.e. information which is not easily quantified or measured. If you are only able to count the number of results which fall into various categories, females or males, voters or non-voters, travellers by bus or travellers by car, etc., this will limit the type of statistical analysis that you can undertake. The various scales of measurement with some examples are as follows.

1. Nominal scales in which only classification and counting are possible. In general, nominal data are not amenable to anything other than simple analysis and do not reveal as much information as data measured on higher scales.

   Higher scales of measurement are as follows.

2. An ordinal scale in which items may be ranked in order of preference or order of occurrence.
3. An interval scale not only ranks items but also gives them a numerical value so that the difference between any two observations can be determined precisely. However, the position of zero on this type of scale is fixed by convention rather than at an absolute zero. Temperature scales are examples of interval measurement as $0°C$ is not a true zero for temperature. So, while the difference in temperature between $10°C$ and $20°C$ is the same as the difference between $20°C$ and $30°C$, a temperature of $20°C$ is not double that of $10°C$.
4. A ratio scale is an interval scale which has an absolute zero such as weight, area, height, etc. This means that ratios such as twice as heavy and so on can be calculated.

Nominal/ordinal/interval/ratio scales go across the continuous/discrete divide. The summary chart in Table 16.1 gives further examples.

Table 16.1 Summary of scales

	Discrete	Continuous
Nominal	Favourite colours	
Ordinal	Preference given as (say) integer 0–10	Preference marked on a continuous 0–10 line
Interval	Adult shoe sizes, British system	Temperature (°F, °C)
Ratio	Number of peas in a pod	Height, weight, etc.

Which Statistical Techniques Can You Use?

Your choice of statistical techniques depends on two factors.

1. The nature of your original problem or questions.
2. The types of data and measurement which you have available.

The following tables are intended to give guidelines as to the suitability of various statistical diagrams and techniques for different types of data and measurement.

Nominal scale

Examples
Voting behaviour
Colour of eyes

Tables and diagrams
Frequency tables
Block graphs
Pie charts
Contingency tables

Sample statistics
Proportions
Estimation of population proportions and confidence intervals

Hypothesis testing
χ^2 test of application (contingency tables)
Tests of proportions

Ordinal scale

Examples
Order of preference
Day of the week (you need to be careful here—does Sunday precede or follow Saturday?)
Ordered categories (e.g. small, medium, large)

Tables and diagrams
Frequency tables
Line graphs
Scatter diagrams for bivariate data
Contingency tables

Sample statistics
Median
Rank correlation coefficient for bivariate data

Hypothesis testing
Significance of rank correlation coefficient
Sign test   } for improvment or deterioration
Rank sign test } following treatment
χ^2 test of association for contingency tables

Counting: discrete data (interval or ratio scales)

Examples
Number of children in family
Number of road accidents in a week

Tables and diagrams
Frequency tables
Line graphs
Pie charts
Stem plots
Box plots
Contingency tables
Scatter diagrams for bivariate data

Sample statistics
Mean
Median, quartiles
Proportions
Rank correlation coefficient
Estimation of population values using confidence intervals
Variance

Possible statistical models
Discrete uniform distribution
Binomial distribution
Poisson distribution
Geometric distribution

Hypothesis testing
Use of binomial model, e.g. can people taste the difference?
χ^2 goodness of fit (for models listed above)
χ^2 contingency tables
Tests on sample means
Tests on sample proportions
Significance of rank correlation coefficient

Continuous data (interval and ratio scales)

Examples
Temperature (interval)
Time of day (interval)
Heights (ratio)
Time taken (ratio)

Tables and diagrams
Frequency tables
Histograms
Stem plots
Box plots
Cumulative frequency curves
Scatter diagrams (for bivariate data)

Sample statistics
Mean, variance
Median, quartiles
Mode
Estimation of population mean, etc., using confidence intervals
Product moment correlation coefficient
Regression lines

Possible statistical models
Normal distribution
Exponential distribution
Continuous uniform distribution

Hypothesis testing
χ^2 for goodness of fit (to models listed above)
Tests on sample means and variances
Paired-sample t tests
Significance of product moment correlation coefficient
Estimation of size of effects and confidence intervals

Interpreting Your Results

What statistical evidence have you obtained? How does this relate to your original problem? What is the purpose of your study? Are you trying to describe a situation? If so, what information have you gathered? Are you testing a hypothesis? If your result is statistically significant, what does this mean? A statistically significant result is not always of practical significance. For example, we found that a sample of 48 packets of Tesco ready-salted crisps had a mean weight which was significantly overweight ($Z = 8$; see Chapter 10). This was because the variability among the content weights of the packets was very small. Thus, as the variance of individual weights was only about 0.06 g, the variance of sample means for samples of size $n = 48$ is 0.06/48. In fact, the sample mean was only 0.28 g higher than the stated average weight, representing about 1 % of the stated contents. The manufacturers will not be unduly worried by this small percentage and in addition will prefer to stay overweight rather than underweight, in order to preserve good customer relations.

The statistical significance of a sample correlation coefficient depends entirely on the size of the sample. For instance $r \approx 0.2$ is statistically significant at the 5% level for samples of about 100 observations. Yet the amount of variance in the data explained by the relationship between the two variables is given by r^2. So in this case, as $r^2 \approx 0.04$, only 4% of the variance is explained by the linear relationship between x and y. Thus the practical significance of this linear relationship is minimal.

Generally, you should take the time to think about the real significance of your results. Can they be generalised to other samples and other situations? Are there other factors in the situation besides those which you have been able to measure? Are there other equally feasible explanations and interpretations which could be placed upon your data?

Writing Up Your Report

Your report writing should follow the stages of planning outlines in this chapter, through to the actual data collection and analysis. Write up your project in such a way that it can be read and understood by the non-expert reader. Both the purpose and the background to the investigation and the statistical techniques used should be presented in simple terms.

Your report should clearly explain the following.

1. What you wanted to find out.
2. What measurements you chose.
3. How you selected your sample.
4. How you collected your measurements.
5. How you analysed your results.
6. What your results mean.

Draw clear tables and diagrams to illustrate your data. Do not include mathematical proofs of formulae and statistical tests you have used. If your tests and results are detailed and complicated, they are better placed in an appendix at the end of the report where the specialist reader may refer to them. The main body of the report should have the results explained in general terms.

In your report, you should also consider the following points. What are the limitations of your study? (Did you have problems in selecting a random sample for instance?) How could your investigation be improved? How could it be followed up in future? Did any unexpected problems arise which gave you results different to those you expected? In general, the main points to remember are be brief, be clear and be 'to the point'!

Guidelines for Designing a Questionnaire

1. Use simple language.
2. Avoid long complicated questions—double negatives, etc.
3. Be unambiguous.
4. Do not ask general questions if you want specific answers.
5. Do not use leading questions.
6. Try to avoid hypothetical questions about situations outside people's direct experience.
7. Be careful with embarrassing questions. Do not make it too difficult for the respondent to admit to socially unacceptable behaviour.
8. Use the minimum number of questions.
9. Pre-coded questions enable you to analyse your replies very easily by computer—but they may force people to give wrong answers.
10. People tend to choose the first response.

11. Ask easy questions first—difficult questions later.
12. *Pre-test* your questionnaire.
13. Do not rely on people's memories over long periods of time.

Non-random Sampling Schemes

Stratified Sampling

Stratified sampling is a modification of random sampling and is used when the population is split into distinguishable groups which are quite different from each other. Examples of this may be age groups, income groups and/or a split in terms of male or female. In such cases a random sample is taken from each group separately and this ensures that the final sample is fully representative. The individual samples may or may not be proportional to the size of the strata. This type of sampling is discussed in Chapter 6.

Cluster Sampling

Cluster sampling reduces fieldwork costs and can be used even when there is no satisfactory sampling frame for the whole population. It involves taking a random sample of large units in the population (for instance local education authorities). Then, from these chosen units, further samples are taken of individuals (e.g. schools in the selected local education authorities). While this type of sampling does have some of the features of random sampling, the investigator does run the risk of obtaining an unrepresentative sample, particularly if there is much variability between the different clusters.

Systematic Sampling

A systematic sample consists of taking every nth, say 10th, member of the population from the sampling frame list (the first sampled item being chosen at random). If the list is arranged in random order, then a systematic sample should approximate to a simple random sample. An alphabetical list is not randomly arranged but may be random as far as the variables under consideration are concerned. This method is often used because it is convenient and does use randomisation. Every item does have an equal chance of being included in the sample.

Quota Sampling

This is a method of stratified sampling in which the selection of members in each stratum is non-random. First, the population is divided into groups in terms of age, sex, social class, etc., and then the number of people in each group to be included in the sample is decided. From then on, the choice of the actual sample units is left to the interviewer.

This type of sampling is used by some national opinion polls (particularly with reference to voting intentions) and also by market research organisations, mainly because it is cheap to administer. When quota sampling is used, any unit in the sample can be replaced by another with the same basic characteristics. This is not so in random sampling, where the interviewer will have to call back when sample members are not at home or otherwise unavailable.

If no sampling frame exists, quota sampling may be the only practical method of selecting a sample. However, if a representative result is to be obtained by this method, then the numbers of people in the population within each stratum must be known so that results can be adjusted in the correct proportions. It must also be remembered that it may be difficult to rely on any statistical results obtained, as the sample is so completely non-random.

Photocopiable Pages

Recording sheet 1
Practical 2.1

Outcome of trial: record H or T

Your total for heads → Total number of heads

Running total for heads → Cumulative total for heads

Trial number	1	2	3	4	5	6	7	8	9	10	11	12	13	14	15	16	17	18	19	20	Total number of heads	Cumulative total for heads	Cumulative proportion of heads
																							$\frac{}{20}=$
																							$\frac{}{40}=$
																							$\frac{}{60}=$
																							$\frac{}{80}=$
																							$\frac{}{100}=$
																							$\frac{}{120}=$
																							$\frac{}{140}=$
																							$\frac{}{160}=$
																							$\frac{}{180}=$
																							$\frac{}{200}=$
																							$\frac{}{220}=$
																							$\frac{}{240}=$
																							$\frac{}{260}=$
																							$\frac{}{280}=$
																							$\frac{}{300}=$

Record
your grand
total

Add your
total to
previous
totals

Recording sheet 2

Practical 2.2

Trial number	Outcome of trial: record number of pins crossing a line					Your total number of pins crossing lines	Running total for number of pins crossing lines	Cumulative probability that a pin crosses a line
	1	2	3	4	5			
								$\dfrac{\qquad}{50} =$
								$\dfrac{\qquad}{100} =$
								$\dfrac{\qquad}{150} =$
								$\dfrac{\qquad}{200} =$
								$\dfrac{\qquad}{250} =$
								$\dfrac{\qquad}{300} =$
								$\dfrac{\qquad}{350} =$
								$\dfrac{\qquad}{400} =$
								$\dfrac{\qquad}{450} =$
								$\dfrac{\qquad}{500} =$
								$\dfrac{\qquad}{550} =$
								$\dfrac{\qquad}{600} =$
								$\dfrac{\qquad}{650} =$
								$\dfrac{\qquad}{700} =$
								$\dfrac{\qquad}{750} =$

Number of throws x	Tally	f	fx
1			
2			
3			
4			
5			
6			
7			
8			
9			
10			
11			
12			
13			
14			
15			
16			
17			
18			
19			
20			
21			
22			
23			
24			
25			
26			
27			
28			
29			
30			
31			
32			
33			
34			
35			
36			
		Σf	Σfx

Recording sheet 3

Practical 2.3

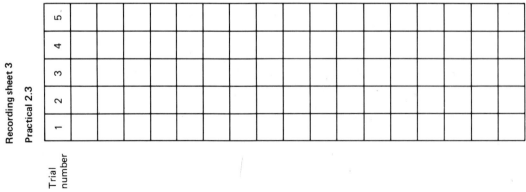

Trial number	1	2	3	4	5

'Push penny' Grid
Practical 2.4

Recording sheet 4

Practical 2.4

Trial number	Outcome of trial: record W or L										Your total for wins	Cumulative total for wins	Cumulative proportion of wins
	1	2	3	4	5	6	7	8	9	10			
													$\frac{}{10}=$
													$\frac{}{20}=$
													$\frac{}{30}=$
													$\frac{}{40}=$
													$\frac{}{50}=$
													$\frac{}{60}=$
													$\frac{}{70}=$
													$\frac{}{80}=$
													$\frac{}{90}=$
													$\frac{}{100}=$
													$\frac{}{110}=$
													$\frac{}{120}=$
													$\frac{}{130}=$
													$\frac{}{140}=$
													$\frac{}{150}=$

Recording sheet 5

Practical 2.5

Top side white

Record the colour of the under side as W or B

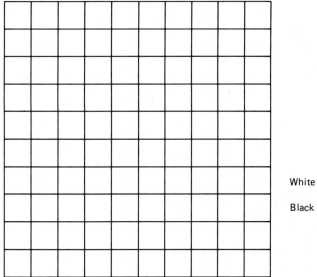

Totals

White

Black

Top side black

Record the colour of the under side as W or B

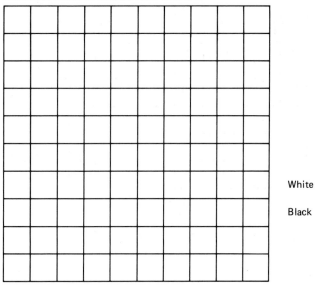

Totals

White

Black

Recording sheet 6

Practicals 2.6a, 2.6b, 2.6c and 2.6d

Die ☐ *	Die ☐ *	Winner

Die ☐ *	Die ☐ *	Winner

Die ☐ *	Die ☐ *	Winner

Number of wins on die ☐ *	
Number of wins on die ☐ *	

*Write in the letters for the dice that you are using: A, B, C and D

PC7

Log Sheet for Collecting Poisson Data

Time interval	Number of vehicles	Time interval	Number of vehicles
1		51	
2		52	
3		53	
4		54	
5		55	
6		56	
7		57	
8		58	
9		59	
10		60	
11		61	
12		62	
13		63	
14		64	
15		65	
16		66	
17		67	
18		68	
19		69	
20		70	
21		71	
22		72	
23		73	
24		74	
25		75	
26		76	
27		77	
28		78	
29		79	
30		80	
31		81	
32		82	
33		83	
34		84	
35		85	
36		86	
37		87	
38		88	
39		89	
40		90	
41		91	
42		92	
43		93	
44		94	
45		95	
46		96	
47		97	
48		98	
49		99	
50		100	

Circles

Blood Cholesterol Levels of 537 Male Patients

Blood cholesterol level									
2.2	1.0	0.6	0.5	1.4	1.5	1.1	1.6	0.9	1.3
0.9	0.5	0.8	0.5	2.5	0.7	1.1	1.3	1.7	3.0
1.8	1.0	1.3	1.1	0.7	3.4	0.9	1.0	0.7	0.8
0.8	1.1	1.3	1.5	0.9	1.4	1.2	0.7	0.8	0.8
1.2	0.5	4.8	0.8	0.7	1.4	0.3	0.6	2.4	0.9
1.4	1.0	2.0	2.5	1.2	0.9	1.5	1.0	3.1	1.5
0.7	0.6	2.2	2.5	0.4	5.1	1.7	3.3	0.8	1.6
1.2	1.2	3.6	2.2	4.5	1.1	1.5	0.9	0.9	1.2
0.6	0.9	2.2	1.2	2.8	2.4	0.8	0.9	1.6	1.2
2.6	0.9	1.6	3.5	2.2	0.7	1.0	1.4	0.7	2.3
1.5	1.5	1.4	1.0	2.0	1.0	1.2	2.4	1.9	0.4
4.6	1.2	1.1	1.7	0.9	0.9	1.5	1.7	1.0	3.2
0.7	1.3	1.2	1.1	0.7	0.6	1.4	1.2	1.1	0.4
2.3	1.2	3.1	1.7	3.0	1.6	2.1	1.1	0.9	3.1
3.3	1.0	1.2	0.4	0.9	1.3	1.0	1.1	2.7	1.2
1.8	1.4	3.4	1.6	1.8	1.2	1.4	1.2	2.4	1.5
1.6	0.9	2.7	0.9	1.6	1.5	1.6	4.9	1.1	0.6
0.7	0.8	0.9	0.5	1.5	4.0	1.0	1.7	1.2	0.5
0.6	2.3	0.9	1.5	1.0	2.9	2.0	1.1	1.5	4.5
1.4	0.5	0.7	3.4	0.5	1.1	1.1	1.1	0.4	0.6
1.0	3.2	2.2	2.2	0.3	1.1	1.3	1.5	0.6	0.9
1.1	2.5	1.9	1.2	1.2	1.0	0.7	0.4	1.1	1.0
3.2	1.1	1.5	1.2	1.4	0.6	1.0	2.3	1.1	4.0
0.4	1.4	0.8	1.4	2.3	0.8	2.2	3.0	1.9	3.8
2.0	2.2	1.1	2.1	0.5	1.0	1.0	2.5	0.7	1.0
1.0	1.4	3.2	2.9	3.0	1.2	1.0	1.8	1.1	2.4
2.0	3.0	3.9	2.7	4.2	2.2	1.6	1.6	1.2	2.0
0.6	2.3	2.0	2.6	1.0	2.0	1.9	1.6	1.6	5.5
1.1	0.7	2.5	2.3	1.7	0.6	1.4	3.2	1.2	1.5
0.7	1.2	2.4	1.2	1.2	2.5	1.8	0.5	1.2	1.2
1.4	1.0	1.4	1.1	2.0	1.0	1.6	1.7	5.4	2.9
0.5	2.3	1.5	1.6	1.5	1.1	5.1	2.8	0.9	0.9
1.6	2.0	2.2	1.4	0.5	1.1	1.0	1.8	4.2	2.4
0.9	1.0	1.6	1.3	3.0	2.4	1.2	2.0	1.2	1.1
1.7	1.4	0.4	0.9	1.2	1.4	1.1	1.2	1.1	1.5
1.9	1.3	1.5	5.3	1.4	1.7	1.3	2.0	0.9	1.4
2.7	0.9	0.5	1.8	0.9	1.6	1.7	1.3	0.5	4.1
1.7	1.5	1.8	1.6	2.0	2.4	1.1	2.5	1.9	0.8
1.2	1.5	1.8	1.9	2.7	2.2	3.0	1.8	1.0	1.5
0.9	1.2	0.8	1.7	1.2	2.0	2.7	1.5	1.6	1.2
1.0	0.8	2.8	1.2	2.0	1.1	5.1	3.6	3.0	1.1
1.4	1.2	2.4	1.4	2.5	0.8	3.0	1.9	1.6	3.6
2.5	1.4	1.2	0.9	2.8	2.5	5.1	0.8	2.4	1.2
2.0	2.0	1.7	1.4	1.4	1.1	0.5	1.8	0.9	1.0
3.2	1.0	2.5	0.8	1.6	1.4	1.5	1.5	2.3	5.0
1.0	1.1	3.2	1.6	0.8	2.0	1.0	1.8	2.5	0.9
0.7	0.7	1.0	1.5	3.8	1.5	2.2	1.2	1.2	1.1
2.0	5.0	0.4	0.9	0.9	2.9	1.0	0.9	1.8	1.1
0.2	2.5	1.0	0.7	6.0	0.8	1.0	1.4	1.2	3.1
1.7	0.6	1.0	0.5	0.8	0.8	0.8	5.5	1.4	0.9
0.9	3.0	1.0	2.0	2.3	2.0	1.2	4.9	3.5	0.6
1.6	2.5	1.2	0.4	0.5	0.8	2.4	1.2	1.7	2.0
1.0	1.9	1.0	0.9	2.0	2.9	3.0	1.2	1.2	1.8
0.5	2.0	0.8	0.9	2.2	1.8	3.6			

Individual Student's Record Sheet for Sampling Exercise

Samples of size n =			Median	Midrange	Mean	Sample variance	
Sample number						Divisor n	Divisor $n-1$
1							
2							
3							
4							
5							
	Total						

Divisor n denoted on many calculators by $(\sigma_n)^2$.
Divisor $n-1$ denoted on many calculators by $(\sigma_{n-1})^2$.

Summary of Pooled Class Results

	Distribution of sample		
	Sample size n =		
	Tally		Number of samples
0.40–0.59 0.60–0.79 0.80–0.99 1.00–1.19 1.20–1.39 1.40–1.59 1.60–1.79 1.80–1.99 2.00–2.19 2.20–2.39 2.40–2.59 2.60–2.79 2.80–2.99 3.00–3.19			

This is to be used for collecting together class results for medians, midranges and means.
Keep samples of size 5 and 10 *separate*.

'Practice Makes Perfect'

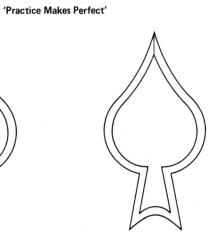

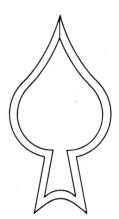

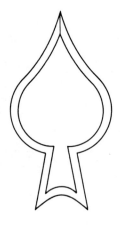
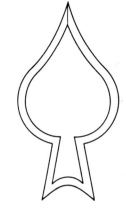

Recording sheet
Simulation 15.3

Start with 1 pair of adult birds

Stage 1 Number of eggs laid.

Three-figure random number   Number of eggs laid

Stage 2 Number of eggs hatched.

Number of eggs laid   Two-figure random number for each egg laid   Number of eggs hatched

Stage 3 Number of chicks surviving to 3 months (fledgelings).

Number of chicks   One-figure random number for each chick   Number of fledgelings

Stage 4 Number of fledgelings surviving to next year.

Number of fledgelings   One-figure random number for each fledgeling   Number of survivors

Stage 5 Number of parent birds surviving to next year.

Toss two coins:

First coin H/T   Number of survivors

Second coin H/T

Recording sheet

Simulation 15.4 (Blue Whales; females; 2 year cycle number ☐)

Age (years)	Numbers at start of cycle	Random numbers for survival	Number of survivors	Random numbers for breeding	Number of calves
0–1				/////	/////
2–3				/////	/////
4–5					
6–7					
8–9					
10–11					
12–13					
14–15					
16–17					
18–19					
20–21					
22–23					
24–25					
26–27					
28–29					
30–31				/////	/////
32–33				/////	/////
34–35				/////	/////
				Total	

Number of female calves, ——.
Number of male calves, ——.
(Add these numbers to the appropriate sheet for the next cycle as the 0–1 years age group.)

PC15

Recording sheet

Simulation 15.4 (Blue Whales; Males; 2 year cycle number ☐)

Add in new calves (males) from previous cycle on females sheet.

Age (years)	Numbers at start of cycle	Random numbers for survival	Number of survivors
0–1			
2–3			
4–5			
6–7			
8–9			
10–11			
12–13			
14–15			
16–17			
18–19			
20–21			
22–23			
24–25			
26–27			
28–29			
30–31			
32–33			
34–35			

Recording sheet 1

Simulation 15.5

Time period	1 Vehicle arrives at M_2	2 Vehicle arrives at S_1	3 Vehicles pass junction from M_2	4 Vehicles pass junction from S_1	5 Length of queue at S_1

Recording sheet 2

Simulation 15.5

Time period	Vehicle arrives at M	Vehicle arrives at S	Lights change	Number of vehicles going through traffic lights		Queue at M	Queue at S
				from M	from S		

Recording sheet 1

Simulation 15.6

Time interval since arrival of previous car			Time taken to buy petrol		
Car number	Random number	Time	Random number	z	Time

Recording sheet 2
Simulation 15.6

Time from start	Arrivals	Departures	Number of pumps vacant	Cars in queue

Index